CW00969389

ALIEN ENCOUNTERS

ALIEN ENCOUNTERS

True-Life Stories of Aliens, UFOs and Other Extra-Terrestrial Phenomena

Rupert Matthews

PICTURE CREDITS:

B. Barber 74, 81, 96, 98, 161, 168; **T. Boyer** 7, 54, 68, 82, 91, 102, 119, 205; **Clip Art** 76; **Corbis** 16, 19, 20, 23, 25, 47, 71, 93, 95, 99, 103, 121, 122, 125, 126, 129, 130, 135, 140, 144, 149, 152, 160, 163, 188, 189, 191, 195, 198; **Getty** 61, 128, 159, 201; **P. Gray** 36, 45, 63, 73, 88, 137; **Kobal** 138, 194; **Mary Evans** 9, 10, 11, 13, 14, 18, 21, 26, 29, 33, 41, 64, 87, 104, 107, 109, 116, 132, 142, 143, 151, 153, 155, 156, 164, 166, 174, 178, 180, 183, 184; **Photos.com** 34, 158; **Shutterstock** 30, 48, 51, 55, 57, 58, 67, 78, 79, 145, 147, 148, 186, 196, 200; **Topfoto** 110, 113

This edition published in 2008 by Arcturus Publishing Limited
26/27 Bickels Yard, 151–153 Bermondsey Street,
London SE1 3HA

ISBN: 978-1-84193-849-3

Printed in China

Contents

Introduction

I first became interested in UFOs during the Great Warminster Mystery, as it was then known, back in the 1960s. I was just a child at the time, but the exciting stories in the press caught my imagination. The idea that aliens from another planet could be visiting our quiet English countryside fired my imagination. We had friends in the area and so I was able to visit the town that was quickly becoming legendary and spend time up on Cradle Hill looking for flying saucers.

I didn't see any, but that did nothing to dampen my enthusiasm. Since then I have grown up, got a job and had a family – but I have never forgotten the excitement of those early days. As the story of *Alien Encounters* has unfolded, I have been a keen enthusiast. I followed the first stirrings of the Roswell incident with interest, I kept abreast of affairs on the UFO front and was amazed by the revelations coming from those who had suffered alien abductions. And like so many other people I dashed to see the movie *Close Encounters of the Third Kind* – great cinema.

And then I decided to sit down one day and think about it all. I trawled back through my newspaper cuttings and library of first hand accounts. I was looking for some clue as to what was really going on. Everyone seemed to have an opinion. Some thought aliens were visiting Earth, others that this was impossible, still more subscribed to the concept that the US Government was guilty of a huge cover-up and a few believed every word the US Government said.

I decided to get right back to the only evidence that counts – the reports given by those who have had an alien encounter. I came at these reports afresh, trying to clear my mind of any preconceived notions or ideas. I looked for features that the reports had in common that might tie them together and give a clue as to the solution to the puzzle.

This book is the result. I have highlighted a number of key alien encounters that teach us something about the phenomenon. I have tried to discuss in a clear and calm way the background to the whole business and to look at what the witnesses tell us happened to them – rather than at what other people have theorized they meant.

The results fascinated me. There is no doubt in my mind that something is going on. There is no doubt that it occurs worldwide and has been going on for a very long time. But what is it?

The evidence is in this book. Read on and draw your own conclusions.

Rupert Matthews

England, 2007

CHAPTER 1

The UFOs Arrive

It is generally believed that the entire alien encounter phenomenon began with the startling sighting of 'flying saucers' or unidentified flying objects (UFOs) by American pilot Kenneth Arnold in 1947 (see Encounter Casebook No. 1 on page 24). Most studies of UFOs ,or books on the subject, feature the Arnold sighting and describe it as the first UFO sighting.

In fact this is very much a matter of using the wisdom of hindsight. At the time, neither Arnold nor anyone else even thought about aliens or UFOs. It was assumed that what he had seen was some kind of top secret military aircraft of revolutionary design. Nor was Arnold's the first sighting of such objects. It was merely the first to make it into the national and international press. For that we must thank the reporter who took Arnold's description of the mysterious aircraft he had seen and dubbed them 'flying saucers'. The name caught the public imagination and made good newspaper copy.

The story took a dramatic new twist when it became clear that whatever Arnold had seen, it was not a secret weapon being developed by the United States Air Force (USAF). The speed, design and motion of Arnold's aircraft were utterly unlike anything being developed at the time. The first thought to spring to the minds of most people in aviation was that the Soviet Russians had developed some startling new technology. However, Arnold's aircraft seemed so far in advance of anything the Russians had used during the Second World War, which had ended only two years earlier, that this seemed rather unlikely.

It was not long before people all across the USA started coming forward with their own sightings of mysterious aircraft. These people had been reluctant to speak publicly before – either because they feared ridicule or because they had not realized that they had seen anything particularly odd: like Arnold they had assumed that they were seeing some secret new type of aircraft.

> It was not long before people all across the USA started coming forward with their own sightings of mysterious aircraft.

It must be remembered that at this time jets, rockets and helicopters were all new inventions that remained shrouded in secrecy and mystery. There seemed to be no limit to the inventiveness of aircraft engineers.

At this early stage the reports that were made to the press or the military were usually fairly vague.

Kenneth Arnold, the American pilot whose sighting of 'flying saucers' made worldwide headlines in 1947. He stands beside the aircraft he was flying at the time of the sighting.

People reported seeing saucer-shaped objects flying very fast, or bright lights at night moving around in unusual ways.

On 19 August 1947, for instance, a Mr and Mrs Busby were sitting on the porch of their house in Butte, Montana, with a neighbour, enjoying the warm evening. A large bright object flew overhead, heading northeast at a tremendous speed. Ten minutes later another ten objects came over flying rather slower, but again heading northeast. As the startled witnesses watched, three of the objects peeled off from the triangular formation and headed due north.

The Busbys did not give any clear description of these objects as regards size, shape or colour. They merely said that they were bright and moved fast.

Somewhat more detailed was the report made by Major Jones of the USAF. This report had great credibility as Jones was the chief intelligence officer of the 28th Bombardment Wing based at Rapid City

Launched in 1948, *Fate* magazine chose to lead its first edition with the Arnold sighting. The cover shows an imaginative re-creation of Arnold's experience, though in reality the flying saucers had been much further away from Arnold's aircraft and were not so disc-shaped as shown here.

A pair of 'foo-fighters' fly alongside a USAF B-24 Liberator bomber over Germany in 1944. The American airmen thought the objects were German secret weapons, the Germans thought they were American devices.

Air Force Base in South Dakota. If an air force intelligence officer was unable to report accurately what he saw in the sky, nobody could.

Jones said that he was walking across the car park at the air base when he saw twelve strange aircraft diving down towards the base from the northwest. The aircraft were in a tight, diamond-shaped formation, indicating to Jones that they were military aircraft. He stopped to watch, wondering what type of aircraft these were.

When the formation was about 6.5 km (4 miles) away, the aircraft began a slow turn at an altitude of around 1,500 m (5,000 ft). Jones could now see that these strange craft were shaped like elliptical discs and each was about 30 m (100 ft) across. Having turned to face southwest, the craft accelerated to a speed estimated to be around 650 kph (400 mph) and climbed out of sight.

It was not only new reports that were surfacing. People were beginning to remember events from previous years which had made no sense at the time, but which now seemed to fit into the flying saucer pattern.

Among these were the 'foo-fighters'. These had been seen in large numbers during 1944, and in smaller numbers in 1943 and 1945. They were glowing balls about 1 m (3 ft) across which flew through the skies over war-torn Europe. The foo-fighters had been seen by the aircrew of Allied bombers on missions over Germany. They were seen flying alongside the bomber formations for several minutes at a time before either disappearing or flying off at high speed.

At first the aircrew thought the foo-fighters to be some form of German weapon or tracking system and tried to shoot them down. But bullets seemed to have no effect on them and since they appeared not to be dangerous, the Allied flyers eventually came to accept them as a feature of the skies over Germany. After the war it was discovered that German pilots had also seen the balls of light accompanying Allied bomber formations. The Germans had taken the objects to be Allied weapons or devices of some kind.

What the foo-fighters actually were has never been discovered.

It soon became clear, however, that the ghost fliers were performing aerobatics and achieving speeds utterly impossible for any known aircraft – and with hindsight impossible even today.

Equally enigmatic had been the so-called 'ghost fliers' of 1930s Scandinavia. These odd aircraft were seen hundreds of times over Finland, Sweden and Norway between 1932 and 1937. When seen in daylight the ghost fliers took the form of extremely large aircraft, bigger than anything then flying, coloured grey and without markings of any kind. At night the aircraft often shone dazzlingly bright searchlights down to the ground. The ghost fliers usually came alone, but sometimes appeared in groups of two or three.

At first the various Scandinavian governments thought that they were being spied upon by top secret scout aircraft from Russia, Germany or Britain. It soon became clear, however, that the ghost fliers were performing aerobatics and achieving speeds utterly impossible for any known aircraft – and with hindsight impossible even today.

Having tried to shoot down the strange intruders, and having spent fruitless days searching for their hidden bases, the Scandinavian authorities lost interest. The sightings faded in 1937 and ceased altogether in 1939 – by which time everyone had more important things on their minds.

In the previous century the years 1896 and 1897 had seen numerous sightings of craft taken to be airships over the USA. These craft were generally around 30 m (100 ft) long and coloured grey without any markings. Sometimes they were seen to have wings or propellers, sometimes not. On a few occasions these strange airships had occupants. These were generally said to be men, sometimes rather short, who spoke some foreign language that could not be understood – though one airship seen over Sacramento in 1896 had a pilot who shouted in apparently good American English: 'We hope to be in San Francisco by tomorrow noon.'

Some of the reports made in the wake of Arnold's sighting were clearly misidentified natural events. A glowing disc-shaped flying object seen over Codroy in Newfoundland turned out to be a meteorite heated red-hot by air friction as it crashed through the atmosphere.

Others, quite clearly, were not. On 15 August 1947 Mr A. C. Urie and his two young sons saw a disc flying low over Snake River Canyon in Idaho while they were on a fishing trip. The object passed them at a distance of just 90 m (300 ft) and though it was moving fast, all three got a good look. It was about 6 m (20 ft) long, 3 m (10 ft) wide and 3 m (10 ft) high. The object had a flange or rim around its base and made a soft whishing noise as it passed. As it flew out of the canyon the object travelled low over a line of poplar trees, which bent and twisted as if caught in a sudden, violent wind.

Potentially more dramatic was the event that occurred on or about 4 July 1947 near Roswell in New Mexico. Radar staff at the Roswell Air Force Base tracked an unknown aircraft flying erratically that suddenly vanished from the radar screens as if it had crashed.

On 8 July the Roswell Air Base press officer, Walter Haut, issued a press release saying that a flying saucer had crashed near the base and that air force personnel were investigating the debris. The press pounced on the story, expecting that the mystery of the flying saucers would soon be solved. Later that same day, however, a rather embarrassed Major Jesse Marcel called a press conference to announce that the crashed flying saucer was, in fact, a weather balloon of a new type that the Roswell men had not recognized.

The Roswell incident would later assume massive importance for those who study UFOs and alien encounters, but in 1947 it was quickly forgotten. Very different indeed was the encounter that took place at Godman Air Force Base in Virginia on 7 January 1948.

Captain Thomas Mantell, the USAF pilot who was killed when chasing a UFO in his P-51 fighter in 1948.

At lunchtime a large, unidentified object had been seen over Fort Knox heading towards Godman Air Base. The guards at Fort Knox phoned the control tower at Godman to report the object. Colonel Guy Hix, commander at Godman, was alerted and

A dramatic artwork showing the moment that Captain Mantell's P-51 Mustang fighter broke up in the air while pursuing a UFO.

sprinted to the control tower just in time to see a large, reddish object fly overhead. The sky was dotted with clouds and nobody had enjoyed a good view of the object.

Hix was not an officer to take chances, however. Something had intruded into the air space that it was his duty to guard, and that something had to be investigated. Hix scrambled the three P-51 fighter aircraft that were kept on standby at the base. The aircraft took off and set out in pursuit of the mysterious object.

It was flight commander Captain Thomas Mantell who sighted the object first as the three aircraft came out of the clouds. 'I've sighted the thing,' Mantell radioed back to Hix at Godman. 'It looks metallic and it's tremendous in size.' A few minutes later Mantell called again. 'The thing's starting to climb,' he reported. 'It's making half my speed. I'll try to close in.'

By this time the two other pilots were being left behind in the chase as both the object and Mantell climbed rapidly. They radioed in to say that they could see the object and Mantell's aircraft giving chase. But then cloud closed in again and they lost sight of the object and Mantell. After searching for a while they turned back to Godman.

Mantell, meanwhile, was in hot pursuit. Ten minutes after his comrades lost sight of him,

Mantell radioed again. 'It is still above me, making my speed or better,' he said. 'I'm going up to 20,000 ft [6,000 m]. If I'm no closer then, I'll abandon the chase.'

After several minutes of radio silence from Mantell, Hix began to worry. Calls were put out, but Mantell did not respond. More aircraft were scrambled to search the skies for the mysterious object and for Mantell's P-51. The aircraft went up as high as 10,000 m (33,000 ft) and spread out for 160 km (100 miles), but they saw nothing.

A few hours later the wreckage of Mantell's aircraft was found strewn over a fairly large area of countryside. It had obviously broken up at high altitude and fallen to the ground in a thousand pieces. The P-51 was a famously robust and reliable fighter that simply did not just fall to pieces for no reason at all.

Suddenly the flying saucer mystery was no longer a bizarre, mysterious talking point to fill newspaper columns on quiet news days. An air force officer had been killed. It was now deadly serious. Questions began to be asked with a new urgency and conclusions were rapidly reached.

Firstly, it was obvious that the USAF would not allow its own officers to be killed chasing after one of their own secret weapons. Whatever the flying saucers were, they were not being produced by the US Government.

Secondly, most people quickly concluded that Soviet Russia would not be testing secret aircraft or missiles over US air space. The risk of one being shot down or crashing was simply too great. If the Soviets did have such aircraft they would surely fly them only over Russian air space, keeping them secret until a war broke out when they could be unleashed to terrible effect.

The flying saucer mystery was no longer just a talking point. An air force officer had been killed. It was now deadly serious.

Thirdly, the flying saucers were generally held to be real objects of awesome power and ability. But what on earth were they?

Various people began to investigate the problem. Some were media reporters, some air professionals, others just interested members of the public. Unknown to the general public, however, the USAF was already undertaking a formal and intensive investigation.

The initial report from Kenneth Arnold, and others that came in during the weeks that followed, were passed to the Air Technical Intelligence Center (ATIC) based at Wright-Patterson Air Force Base in Ohio. ATIC was the unit charged with assessing the aircraft and air weapons of other countries, so it was natural that any reports of strange aircraft would be sent there for analysis.

The staff at ATIC were baffled. They were accustomed to studying photos, technical data and other intelligence reports detailing the performance of aircraft not very different from those flown by the

A F/A-18C Hornet fighter flies over the Pentagon. The US military authorities based at the Pentagon were at first seriously alarmed by UFO reports and made strenuous efforts to find out what was happening.

USAF. They were now being asked to deal with eyewitness reports of craft with seemingly highly advanced abilities. A decision was made to ignore reports from civilians, but to chase up and investigate any reports that came in from air force personnel.

The military top brass at the Pentagon were pushing ATIC for a quick and definitive answer. All that ATIC could report was that they were working on the problem. In September 1947, ATIC produced a carefully worded interim report that avoided giving any firm conclusions. The basic thrust of the report was that many, perhaps all, reports could be explained as natural phenomena of some kind but that a small number of reports probably could not be dismissed so lightly. These few reports, ATIC stated, indicated that the 'flying discs are something real and not visionary. There are objects approximating the shape of a disc and of such appreciable size as to appear to be as large as a man-made aircraft.'

The senior officers at the Pentagon were unimpressed by the vague and inconclusive report. They were responsible for the air defence of the USA and wanted not only a quick and definitive answer to what people were seeing, but also a clear course of action that they could in turn recommend to the politicians.

The situation was not very much advanced when Captain Mantell was killed. Colonel Hix was a well-known and highly regarded officer, and Mantell was considered to be a crack pilot. The Pentagon was seriously alarmed that the USA now faced a real, deadly and immediate threat. The USAF was ordered to get to the bottom of the problem, and set up Project Sign. This was an investigative team of air force officers and civilian experts tasked with going over all known reports of flying saucers to produce a final and definitive answer.

Among the civilian experts recruited to Project Sign was Dr J. Allen Hynek. Hynek was an astronomer of great repute whose task was to see if any of the reports could be explained as being meteors, comets or stars seen under unusual conditions. Hynek would later become one of the most respected UFO investigators in the world, but at this stage was simply an astronomer hired to do a job.

Project Sign investigated 237 sightings, selecting them on the grounds of witness credibility and detail of observation. After a year of work, Sign concluded that 77 per cent of these reports could be explained as comets, stars, conventional aircraft, weather balloons or other perfectly normal things seen under unusual or misleading circumstances. The remaining 23 per cent were put down as 'Unknown explanation'.

For public consumption a new Project Grudge was created. This produced a report that was released to the public. It was basically the Project Sign report, but with any sensitive or secret information edited out. The Grudge Report, as it became known, was issued on 27 December 1949. This played down the results, and recommended that the United States Air Force ceased detailed investigations of flying saucers.

With some 23 per cent of reports being classified as 'unknown', the Grudge recommendations might be considered odd. From the point of view of the USAF, however, they made perfect sense. The task of the air force was not, and is not, to investigate strange phenomena no matter how interesting they might be. Its task is to protect the USA and its allies from air attack.

By 1949 the cold war with the Soviet Union was becoming serious. The Soviet Union had many hundreds of jet fighters and jet bombers, plus an

The cover of the seminal book by Major Donald Keyhoe in which he first advanced the theory that flying saucers were alien spacecraft, and that the US military was covering up the truth.

unknown number of rocket powered missiles for the USAF to worry about. Whatever the flying saucers were, they were not Russian, possessed no obvious weaponry and posed no clear or immediate threat.

The first worries caused by the death of Captain Mantell had by this date been allayed. An investigation had found that Mantell's P-51 had not carried the usual high-altitude oxygen equipment. This was probably why Mantell had said he would halt his pursuit at 6,000 m (20,000 ft). But in the circumstances he may have been tempted to go higher. If he had done, he may have blacked out due to lack of oxygen.

The P-51 would then have continued its climb until it stalled, then fallen nose down into a dive accelerated by its engine. Within a few seconds the stresses of a steep power dive would have torn the wings off the fighter and caused it to disintegrate. Satisfied that Mantell had died due to an accident, the USAF concluded that the incident did not prove that flying saucers were dangerous.

With many pressing demands on its time, manpower and budgets, the USAF decided to sideline the whole flying saucer business. It was not the task of the USAF to investigate anomalous phenomena. Its prime job was to protect the USA and its allies from air attack, and the flying saucers did not pose a threat. Any reports that did come in were still passed to ATIC, but they were simply filed without investigation.

The press and the public had not lost interest. Reports continued to be made and were featured by the press and broadcast media. Some reporters took a particular interest in flying saucers, as did some air professionals and members of the public.

Since it was now becoming clear that only some of the strange aircraft were disc-shaped – Arnold's originals had been crescent-shaped – these researchers began to drop the popular name of flying saucers. Instead they adopted the term unidentified flying object (UFO). This was thought to

An early atomic bomb test carried out by the US military at Bikini Atoll. The advent of atomic power was widely thought to have attracted the attention of aliens to the developing human civilization.

lend an air of scientific rigour to the various investigations and meant that the investigators were not prejudging the eventual solution to the riddle.

Among those still working on the problem was retired US Marines officer Major Donald Keyhoe. Keyhoe had extensive personal contacts within the military. These enabled him to talk to many witnesses first hand and to see many written reports, and although he was denied access to any of the secret reports he did learn of their existence. Keyhoe gradually came to a series of startling conclusions, which he published in his classic 1950 book *The Flying Saucers are Real.*

Keyhoe believed that Earth had been under observation by beings from another planet, or series of planets, for some 200 years or more. He believed that the sudden upsurge in sightings after 1947 was linked to the detonation of the first atomic bomb in 1945. Keyhoe argued that the invention of atomic power was a key moment in the progress of a

A group of senior USAF technical officers which met in 1952 to review the evidence relating to UFOs. Captain Edward Ruppelt, who was to lead Project Blue Book and so be in charge of the official investigation into UFO activity, is standing third from left.

civilization and that any more advanced culture would be bound to take notice of the crossing of this threshold. The increased sightings of flying saucers were, Keyhoe argued, evidence that the previously sporadic observation of humanity by aliens had been stepped up. In due course, Keyhoe argued, the aliens would make contact – though when, where or how would be up to them.

Keyhoe's final conclusion was that the Grudge Report had been the public facade of a top secret report. Keyhoe believed that the hidden secret report had concluded that the flying saucers were alien spaceships but that releasing that information to the public would cause such panic that the information had been suppressed.

Keyhoe's book was an immediate and staggering success. It was reprinted just two weeks after being published and eventually sold over half a million copies. By the autumn of 1950 the idea that flying saucers were alien spacecraft had taken a firm hold of the public imagination and was assumed as fact in most press reports of flying saucers.

In 1952 the USAF decided that the continuing number of reports coming in of flying saucers and other unknown objects demanded some sort of response. In March Captain Edward J. Ruppelt was put in charge of the by now bulging files at ATIC. His task was to respond to inquiries from the press or public and to try to bring some sort of order to the files and their contents. Ruppelt was given neither orders nor resources to investigate reports or theories – the USAF believed that task had been completed with the Grudge Report.

Ruppelt turned to various officers and others, including Hynek, who had an interest in the subject and persuaded several to work part-time or without pay on the new initiative. Ruppelt's task was given the codename Project Blue Book.

Just four months later the flying saucer story took a dramatic new turn, when a formation of the mysterious craft flew over Washington, DC, the capital of the USA. On 19 July 1952 a dozen or more saucers flew over central Washington, including the White House and Capitol. The objects were tracked on radar and jet fighters sent up to intercept, but they had vanished by the time the jets arrived. The UFOs were back on 26 July, again flying over the very heart of US power and prestige.

The dramatic sightings not only put flying saucers back at the top of the news agenda, they also created a widespread alarm that aliens were about to invade. Hundreds of reports of new sightings of discs and other flying objects poured into the USAF and were forwarded to Captain Ruppelt at Project Blue Book. Unable to cope with the mass of paperwork deluging his office, Ruppelt appealed for help. In response the USAF asked the Battelle Memorial Institute, a leading scientific research facility located in Ohio, to undertake a study of the most recent reports.

Meanwhile, the Central Intelligence Agency (CIA) decided to take a hand. The CIA had become seriously alarmed by the saucers seen over Washington and by the vast number of subsequent reports that summer. It was not that the CIA thought that Earth was about to be invaded by aliens, nor even that flying saucers were anything other than natural objects or hallucinations. The CIA were worried about the Russians.

In 1976 the archives of Project Blue Book began to be declassified and made available to the public. However, much of the more sensitive information was kept secret for another 30 years and even now not all the files are available uncensored.

If the Soviets were ever going to launch an air attack on mainland USA they were hardly going to phone up the White House and announce the fact beforehand. The first the Americans would know of the attack would be when missiles tipped with atomic warheads came streaking through the skies. If those missiles were mistaken for flying saucers then they might well land and explode without the USAF having taken any action to intercept them.

The CIA concluded that the best action was to persuade the American public that flying saucers simply did not exist.

Just as great a worry for the CIA was the flood of reports that followed the Washington sightings. These overwhelmed not just Project Blue Book, but most of the reporting systems of the USAF designed to watch for intruding enemy aircraft and missiles. The CIA reasoned that the Soviets might instruct their agents in the USA – and nobody knew precisely how many of these there were – to release balloons, fly model aircraft and otherwise start a rash of flying saucer reports in order to swamp the USAF reporting systems as cover for a real missile attack.

The CIA concluded that the best action was to persuade the American public that flying saucers simply did not exist. To do this they needed to establish firmly in the public mind that all reported flying saucers were perfectly normal objects that had been misperceived by witnesses. The CIA therefore persuaded Dr H. P. Robertson, a highly respected physicist, to convene a panel of scientists who would study saucer reports and pronounce them to be prosaic and entirely innocent objects.

The Robertson Panel reported in January 1953. As wished by the CIA, the report concluded that unidentified flying objects were just that – quite normal objects that for one reason or other were unidentified by the witnesses who saw them. It recommended that active steps be taken by the government to educate the public of this conclusion, and further that military staff should concentrate on genuine hostile powers rather than theoretical aliens.

In short the Robertson Panel recommended that the US Government should actively dismiss and debunk all UFO reports. This was to become the policy – official or not – of the US Government, the USAF and its Project Blue Book and other government agencies. The earlier disinterest was replaced by active hostility. The extent to which Project Blue Book, the FBI, the CIA and others have pursued this policy and the techniques they have used is a matter of ongoing controversy.

In 1969 the USAF announced that 'the continuation of Project Blue Book cannot be justified either on the grounds of national security or in the interest of science.' Project Blue Book was closed down. Officially the US Government has had no further interest in the UFO or alien issue. Many

President Richard Nixon meets with top aides in the White House in 1969. By this date the US Government had lost interest in the UFO phenomenon – at least officially.

people suspect that it has had a very active unofficial involvement, but if this is true the government have shrouded the issue in mystery and secrecy.

What is beyond doubt is that the outright denial by the US Government that UFOs existed at all bred distrust and hostility among those who were investigating the phenomenon. Many suspect that behind the activities of the Grudge Report and the Robertson Panel there lurked a darker and more ominous reality. It is argued that if the US Government did in fact know that UFOs were alien spacecraft, and perhaps had even captured one, then the most natural course for them to take was to deny that UFOs existed so that they could keep the secrets that they had gained. Under this theory, all government activity and announcements have been designed to cover up the truth.

By early in 1953 the scene was set for the following decades of alien encounters. Most governments, following the US lead, have either denied that UFOs exist or remained aloof from the subject. Meanwhile members of the public have continued to see and experience strange, bizarre and sometimes terrifying things. It has been left to part-time amateur investigators to try to make sense of what is happening.

Encounter Casebook No. 1

TYPE>>**Daylight disc** DATE>>**24 June 1947** PLACE>>**Mount Rainier, Washington State, USA** WITNESS>>**Kenneth Arnold**

Kenneth Arnold's 1947 encounter is widely regarded as the 'original' UFO sighting: although it is highly likely that there had been sightings before, his was the first to make headlines and to draw international attention.

Kenneth Arnold was an experienced pilot who had been in Chehalis, Washington State, on business. On the morning of 24 June he set off to fly home to Oregon in his single-engine Callier light aircraft. With both time and fuel to spare he decided to first spend an hour or so over the Mount Rainier area searching for a US military transport aircraft that had been reported missing and was presumed to have crashed.

It was while turning on to a new leg of his search pattern at an altitude of 2,800 m (9,200 ft) that Arnold saw a bright flash of light sweep over his aircraft. Such a thing usually happened when sun reflected off the surfaces of another aircraft close by. Fearing a collision with an aircraft he had not seen, Arnold hurriedly levelled his plane and scanned the skies, desperately seeking another aircraft. He soon saw a DC4 airliner some miles distant and flying away from him. Discounting this as the source of the flash, he then saw a second flash far to the north.

Staring at the location of the flash, Arnold saw a line of nine aircraft flying towards him at an angle. As the aircraft came closer he saw that they were flying in echelon, a usual military formation, but arranged with the lead aircraft above the others contrary to the standard military practice. Arnold at this point assumed that the fast-approaching aircraft were military jets of some kind and relaxed.

But as the nine aircraft came closer, Arnold was able to see them in detail and at once realized that he was seeing something very strange indeed. Each aircraft was shaped like a wide crescent with neither fuselage nor tail. Moreover the aircraft were flying with a strange undulating motion quite unlike the straight-line flight of all known aircraft. They also fluttered or dipped from side to side at times, sending off bright flashes as the sun reflected from their highly polished silver-blue surfaces. There were no markings that Arnold could see, though he was now concentrating hard on the mysterious aircraft.

> ... the aircraft were flying with a strange undulating motion quite unlike the straight-line flight of all known aircraft.

A flight of PT-21 USAF training aircraft in echelon formation. Arnold at first thought that the UFOs were military aircraft as they were flying in a formation similar to this.

Signed by Kenneth Arnold, this artwork shows the aircraft that Arnold saw during his famous sighting. The crescent shape and smooth metallic surface would later be reported scores of times by other witnesses.

The formation was moving fast. Arnold timed it as it passed over landmarks on the ground and later estimated the speed at around 2,100 kph (1,300 mph). This was much faster than any known aircraft of the time. Even military fast jets flew at only around 1,100 kph (700 mph). The aircraft were soon out of sight.

Arnold had begun to worry that he had seen some sort of highly advanced Soviet war machine.

Arnold headed for Yakima Airfield and went to see Al Baxter, the general manager of Central Aircraft. The two men discussed the sighting, and Arnold drew pictures of what he had seen. Other pilots and air crew joined the conversation, but none could explain what Arnold had seen, other than to guess that the strange aircraft were some kind of secret military project. Still confused, Arnold then resumed his interrupted flight home to Oregon.

By the time he arrived, Arnold had begun to worry that he had seen some sort of highly advanced Soviet war machine. He decided to inform the FBI, but their office was closed, so he dropped in at the offices of the *East Oregonian* newspaper. He told the reporters there all about his experience. One of them, Bill Bacquette, queried the way the unusual craft moved. Arnold elaborated on the undulating motion by saying, 'They flew like a saucer would if you skipped it across water.'

Bacquette filed his report with a national news agency, writing about 'flying saucers'. It was repeated across America and soon the public was agog at news of these flying saucers.

Meanwhile, Arnold had returned to the FBI to tell them about the strange aircraft. The local FBI man passed the details on to head office in Washington concluding his report with the words, 'It is the personal opinion of the interviewer that Arnold actually saw what he states he saw in the attached report.' Already concerned about Russian intentions and military technology, the US military pounced on Arnold's report.

An era was born.

CHAPTER 2

Out of This World

In the years since the US Government publicly turned its back on investigating the UFO phenomenon, the study of the mystery has been largely in the hands of journalists, amateur enthusiasts and a few full-time investigators.

The quality of the investigative work undertaken has been mixed, with some pursuing open-minded strategies that would do credit to any scientific institution while others adopt a highly superficial approach. Some have approached the problems posed by UFOs with quite open minds; others seem to have decided in advance on one particular solution to the mystery. Among what might be called 'ufobelievers', the most common fault is to try to adapt all evidence to fit the theory that aliens are visiting earth, while among 'ufosceptics' the determination to establish that all UFOs are natural phenomena can lead to evidence being ignored or to witnesses being accused of lying or living fantasy lives.

> Hynek had become convinced that UFOs were real, though he studiously kept an open mind as to what they were.

Whatever their viewpoint and methods, however, nearly all UFO investigators have come to agree on certain key things. Perhaps the most important of these is the classification of UFO sightings into six or more categories with generally agreed names and meanings. The classification system was developed in 1972 by Dr J. Allen Hynek, the astronomer who had formerly helped with Project Sign and Project Blue Book.

When Hynek was initially called in by the USAF he was a sceptic, thinking that UFO witnesses were mistaken or fraudulent in what they reported. He gradually came to change his mind, however, as he gathered data and interviewed witnesses. By the late 1960s Hynek had become convinced that UFOs were real, though he studiously kept an open mind as to what they were. He died in 1986 after setting up the Center for UFO Studies, generally referred to as CUFOS.

Apart from Hynek, serious scientists have generally steered clear of UFO studies. In general this is because they believe, as Hynek did initially, that UFO witnesses are either mistaken in what they report or are not telling the truth. They tend to follow the line of reasoning most publicly followed by Project Blue Book: that UFOs cannot exist according to known laws of physics and so therefore they don't exist.

Professor Allen Hynek was at first sceptical about UFOs, but the more he studied the details of reported sightings, the more convinced he became of their reality.

Scientists are also human beings, and there is general fear of ridicule from colleagues. Moreover, openly supporting a theory or claim that is not only against the consensus, but is later proved to be false, can be harmful to a career in science. More practically there are virtually no funds available from either governments or universities to support research into UFOs. Since scientists need employment to pay the bills and put food on the table they can hardly be blamed for preferring to study more mainstream subjects that supply them with paid jobs.

An artist's impression of a trio of UFOs over mountainous terrain. Close encounters make up a minority of reported sightings, but they alone provide the sort of detail seen in this view.

Hynek's first three categories are collectively known as 'encounters' and generally refer to UFOs seen at a distance. The first of this trio is the 'nocturnal light' which occurs when a witness sees a light in the sky at night behaving in a way that cannot be explained rationally. The second category is that of 'daylight disc', which includes all UFOs seen in the sky during daylight whether they are actually disc-shaped or not. The final agreed encounter category is the 'radar contact', which refers to UFOs picked up on radar. Some use a fourth encounter category: 'radar-visual' refers to UFOs that are tracked on radar at the same time as being seen, whether by day or night.

The second group of categories are 'close encounters', sometimes referred to as CEs. A 'close encounter of the first kind' is when a UFO is seen at close quarters and for a fairly prolonged period of time. The witness is able to give a detailed description of the UFO, its shape, colour and behaviour.

A 'close encounter of the second kind' is essentially similar to a CE1, but where the UFO has some clear impact on its surroundings. This might be as simple as causing vegetation to sway as it passes, or may involve burning plants or ground with what appear to be engine blasts. A key feature is that the effect must be unique. To take an example, if a UFO is seen to land and marks are afterwards found where the UFO rested these marks should be unique and not identical to marks left by farm machinery nearby. Some UFO investigators collect samples of burnt grass and disturbed soil in the hope that analysis may reveal something about the motive power or composition of the UFO that affected them.

A 'close encounter of the third kind' occurs when a CE1 or CE2 is combined with the appearance of what seem to be occupants or crew from the UFO. These beings are sometimes referred to as 'ufonauts', a fairly neutral term that is used in an effort to avoid prejudging who or what these beings might be. Most of those who have a close encounter of the third kind are fairly convinced that they have met aliens of some kind or another.

These aliens come in many different shapes and sizes. Their behaviour can range from mundane to bizarre through various degrees of friendliness and outright hostility. As we shall see, however, there are some features and strands that seem to link a number of cases and may form the basis for understanding what is going on.

Some researchers include other CE categories. A 'close encounter of the fourth kind', for instance, may be considered to be one in which the witness interacts with the alien in some way. Some classify say that CE4 is where the witness and the alien communicate. Others think a CE4 should refer only to those incidents where the witness is abducted by the aliens. A 'close encounter of the fifth kind', or according to some of the sixth kind, occurs when a witness has repeated and ongoing contact with the same alien.

Their behaviour can range from mundane to bizarre through various degrees of friendliness and outright hostility.

Of the close encounters, only CE1, CE2 and CE3 are universally accepted. The other categories of CE are not accepted by all researchers, and can cause some confusion as not all who do recognize them see them as meaning the same thing.

The vast majority of UFO reports, however, belong not to the high profile CE2 and CE3 categories, but to the less well known but equally valid CE1 and encounter categories. Some of the most famous UFO sightings, including that by Kenneth Arnold (see Encounter Casebook No. 1 on page 24) belong here. Today Arnold's sighting would be categorized as a daylight disc and would probably attract little attention.

Although the sightings of the encounter and CE1 categories may not seem to be overly exciting when viewed individually, they make up an impressive body of evidence when taken together. There are certain features that are repeated time and again in different reports separated by great distances of time and space. Taken together these may indicate what sort of object is being encountered.

One of the earliest reports came from Baradine in New South Wales, Australia. It was made in 1948 but referred to events in 1931 when a group of boys was out rabbiting by moonlight. The boys were intent on setting snares, but one noticed that he was casting two shadows. Looking up, he saw a disc-shaped object as bright as the moon approaching from the northwest. What looked like orange lights or flames flashed on and off around its rim and the object could be seen to be rotating slowly as it moved. The disc flew in a straight course until it disappeared behind nearby hills. The boys told their families what they had seen, but nobody took any notice until after Arnold's sighting when the story got into the local press.

In January 1958 a Brazilian survey ship, the *Almirante Saldanha,* arrived at the Pacific island of Trindade, where the Brazilian Navy had an oceanographic station. Just after noon on 16 January the photographer Almiro Barauna was on deck when another crew member pointed out to him an object in the sky. The object was at first taken to be an aircraft, but its lack of wings made Barauna reach for his camera. The object circled around the island, then flew off. Barauna managed to take four photos, though in his excitement he did not check the settings on the camera and all four pictures were consequently slightly over-exposed. About a hundred sailors and residents at the base saw the UFO.

Realizing that there must be no suspicion of fraud or hoaxing, Barauna persuaded the captain of the ship to supervise the developing of the photos in the on-board laboratory and on the ship's return to port he submitted the negatives and prints to the Brazilian Navy for expert study.

The developed photos matched exactly the descriptions given by the witnesses. The craft was shown as being a flattened sphere with a wide rim or flange around its centre, giving a rather Saturn-like appearance. The body of the craft was pale grey, the rim dark grey and a greenish mist or spray trailed behind it. The overall diameter of the craft was estimated to be about 40 m (130 ft) and its speed around 950 kph (600 mph).

The object was at first taken to be an aircraft, but its lack of wings made Barauna reach for his camera.

The Trindade sighting is famous largely because of the photos that were taken. The UFO shows features that are repeated in many sightings. The shape of a flattened sphere with a rim is one that is reported in a great many cases. That apart the

One of the famous photos taken by Almiro Barauna of the UFO that was sighted from the Brazilian ship *Almirante Saldanha* in 1958. The rounded shape and horizontal flange are typical of many sightings.

behaviour of the UFO was not particularly noteworthy. It flew at a speed easily attained by conventional aircraft and its flight path, circling the island and heading off in a straight line, could be mimicked by a human craft. Were it not for the photos, sceptics might have dismissed this sighting as being that of a misidentified aircraft.

No such mistake was possible in the case of the UFO that 'buzzed' a passenger jet just before midnight on 11 November 1979. The Supercaravelle twin engined jet liner of the Spanish airline TAE was flying from Salzburg to Tenerife when at around 11 pm the pilot, Lerdo de Tejada, saw two bright red lights off to port and at about the same altitude as his own aircraft, 7,500 m (24,000 ft).

As the two red lights gradually approached, Tejada realized that they were fixed to a rather larger flying object. Due to the dark night and dark colour of the object, Tejada could not get a clear idea of how large the thing was nor its shape. Following

A twin-engined civilian jet takes off at dusk. The classic Tejada sighting of 1979 was made by the pilot of such an aircraft.

international procedure, Tejada called up Barcelona Air Control, in whose area he was, to report that a second aircraft was close to his. Barcelona confirmed that they had a second aircraft on their radar screens but that they could not identify it.

Tejada decided to alter course to stay clear of the intruder, then return to his path towards Tenerife. As he began the turn, however, the two red lights suddenly accelerated and climbed. They were now about 750 m (2,500 ft) from the airliner and began a bizarre dance around the TAE plane. One minute they were following the airliner, then the next they were immediately above it, then they swooped down below. Tejada was by now alarmed. The strange aircraft seemed to be playing games with him, and very dangerous games at that. When the red lights swooped down into the path of the Supercaravelle, Tejada had to haul his aircraft round in a sudden, tight turn to avoid a collision.

Tejada at once radioed Air Control and demanded permission for an emergency landing at the nearest available airport. He was directed to Valencia Airport, which was alerted to the approach of not only the jet but also the unidentified intruder. The TAE aircraft was followed to the airport by the UFO. As Tejada landed, the staff at the airport confirmed seeing it followed down by a large craft with two red lights. Seen from the ground the UFO was estimated to be as large as a Boeing 747. Once the passenger jet was on the ground, the UFO streaked off at high speed and vanished from radar screens.

Although the UFO was not seen clearly, the sighting does exhibit some interesting features. The UFO was seen by several people and was picked up on radar, so it can be assumed that it was a real object and not merely a star misinterpreted by Tejada.

The UFO was obviously attracted to the passenger jet. When first seen it was some distance away, but closed in rapidly and thereafter stayed close to the jet until it landed. This could suggest that the craft showing the two red lights was under some sort of intelligent control and may have been investigating the aircraft in some way.

Another Supercaravelle passenger jet was involved in a sighting over Brazil on 7 May 1967. The pilot and co-pilot both spotted a disc-shaped aircraft ahead of them as they approached Porto Alegre. The disc was off-white in colour with a row of flashing red lights around its rim. The aircraft's course would take it past the object with plenty of room to spare, but the co-pilot kept the disc in view. As the airliner drew level with it the disc began to move, falling in beside the aircraft at a distance of about 500 m (1,600 ft) and keeping pace with it. After about 20 minutes the object altered course, accelerated sharply and disappeared into the distance.

A similar but more dramatic sighting took place over Suffolk, England, on the night of 13 August 1956. The action took place around two Royal Air Force (RAF) bases that in the 1950s were leased to the USAF: RAF Bentwaters and RAF Lakenheath.

The pilot and co-pilot both spotted a disc-shaped aircraft ahead of them as they approached Porto Alegre.

At 10.55 pm the radar operator at Bentwaters picked up an unidentified aircraft approaching fast and low from the east. The pilot of a C47 transport plane that was in the area was alerted and asked if he could see anything. The pilot looked down and saw a large, soft light flash past underneath him at an estimated 3,200 kph (2,000 mph). The light was seen simultaneously by ground crew at Bentwaters. Because the object was heading for Lakenheath, that air base was alerted by phone.

Soon after, Lakenheath spotted the object approaching from Bentwaters, as well as two others. These three objects proceeded to perform some amazing manoeuvres quite beyond the abilities of any earthly aircraft. They moved at high speed, stopped dead and turned at right angles. Not only were such stunts beyond any known aircraft, but any human inside a craft turning so sharply would have been knocked unconscious and possibly killed by the forces involved.

A dramatic re-creation of the dogfight over RAF Bentwaters between a UFO and an RAF jet fighter in 1956.

Worried, the USAF radar operators at Lakenheath called up the RAF, which scrambled a Venom jet fighter from RAF Waterbeach to investigate. The Venom was guided towards the strange objects by ground radar, then picked them up on its air-to-air radar. The Venom closed in to get a clear visual sighting, but the UFO suddenly accelerated, climbed and then dived down to get on to the tail of it. Now the hunted not the hunter, the Venom pilot threw his aircraft around in desperate attempts to get the UFO off his tail, but to no avail. After several seconds of fast-moving aerial action, the UFO abandoned the pursuit and took off east at high speed.

The fact that the sighting showed up on radar and was made by trained military personnel accustomed to seeing all sorts of objects at night adds to the credibility of the event.

Although the Bentwaters UFO was never seen clearly, all who saw it agreed that it was about the size of a fighter aircraft and glowed with a soft, pearly light that appeared fuzzy or indistinct at the edges. Again, it appeared to be controlled by some intelligence, especially during the dogfight with the Venom. The fact that the sighting showed up on radar and was made by trained military personnel accustomed to seeing all sorts of objects at night adds to the credibility of the event.

As with the Tejada sighting, the Bentwaters UFO acted as if it were being controlled by some intelligent being. It was at first flying on a course of its own, but then became attracted to the aircraft. Even more noticeably than the Tejada UFO, that sighted over Bentwaters performed manoeuvres that would be quite impossible for any known aircraft. Indeed, so sudden and dramatic were the changes of speed and direction that any human inside the craft would have been seriously injured or killed.

In 1957 a close encounter of the first kind took place over New Jersey that shows other features of UFOs. A lady in Trenton who preferred to remain anonymous was clearing up her back room at about 2 pm on 6 March when she heard her dog barking in the backyard. Going out to see what was causing the fuss she saw that the dog was barking excitedly while looking upwards. Looking up herself the woman saw a round flying object some 150 m (500 ft) away and about 15 m (50 ft) across. The woman likened its shape to that of a derby or bowler hat. The central domed area was about 9 m (30 ft) high with steep sides while the flat bottom extended beyond the dome to form a rim about 5 m (15 ft) wide. The colour and texture of the object was that of pipe clay, a smooth but dull off-white substance.

As the woman watched, the object began to rock or sway slightly from side to side. A low rumbling noise began that grew louder, then faded only to become louder again. There then came a soft whooshing noise and the object rose vertically to disappear into the clouds.

The woman immediately phoned her husband, who was at work in New York. He advised her to write down an account while it was still fresh in her mind. This she did, later sending the account to the USAF where it was quietly filed away in Project Blue Book.

The object seen over Trenton continued the same description of a round shape with a rim reported elsewhere. Because it was seen at close quarters in daylight details of its movement could be made out. The wobbling or rocking motion is one that has been seen in many UFOs.

> The glowing object pulsated with various colours bright enough to light up the surrounding area.

Many sightings, while convincing and apparently genuine, do not add much detail to our knowledge of UFOs. One evening in 2001 George Sykes, a retired hospital worker living in a small village outside Aberdeen in Scotland, spotted two large spherical craft flying past his house. The globes were spinning as they flew and glowed with a soft light that changed colour every few seconds. They flashed past then vanished into the evening air.

In 1980 bus conductor Russel Callaghan was on his usual route through the Yorkshire countryside near Bradford. Having reached the end of the journey, Callaghan and his driver halted the bus for a few minutes before the return trip. As they sat having a cigarette on the grass the two men spotted a silver disc hovering over Emley Moor a few hundred metres away. The disc began to spin, gaining speed and then began to move. It gathered speed quickly and streaked out of view in about 8 seconds.

Rather more precise in their descriptions of the object seen were Deputy Sheriffs Dale Spaur and Wilbur Neff of Portage County, Ohio, who had a close encounter of the first kind on the night of 16 April 1966. The two deputies were on patrol on Route 224 when they stopped to investigate an abandoned car on the roadside.

While Neff waited beside the police car, Spaur walked over to the wreck. Spaur inspected the car, concluded that nobody was about and turned to return to the police car. He then saw a large, brightly lit object coming towards them. Spaur told Neff to turn around, which he did before freezing in alarm. The object was meanwhile getting closer, and could now be seen to be about 12 m (40 ft) tall. It was emitting a quiet humming sound. The glowing object pulsated with various colours bright enough to light up the surrounding area. It was shaped like an American, or rugby, football standing upright though the top was more rounded and domed than the more pointed bottom.

The object hovered over the police car for a few minutes at a height of about 250 m (800 ft), then moved off east. Spaur and Neff gave chase in their car, being later joined by Officer Wayne Houston

US police officers, who are trained to note details and appearances, have proved to be valuable witnesses when they encounter UFOs in the course of their duties.

who was parked up on Route 14 when the UFO came flashing past, chased by Spaur and Neff. The object halted again near the village of Harmony. The three policemen stopped and got out of their cars to watch the UFO as it hung in the air, then climbed up at speed and disappeared from view.

The key point about the Portage County sighting is that the object was seen at close quarters and from various angles over a considerable period of time by three different witnesses. There seems little likelihood that they mistook a star or other object for a UFO. As with so many other UFOs, this one was rounded in shape, glowed with coloured lights and was able to move at speed.

One witness who acted with commendable coolness when confronted with an unknown object at very close quarters was Australian farmer A. Pool, who was driving back to his farm after a long day tending sheep on his station in Western Australia. Pool was trundling across a grassy paddock in his off-road vehicle when he spotted an aircraft heading towards him. Thinking a pilot was in trouble, he braked to a halt to await events. The approaching aircraft turned out to be a grey-coloured disc flying at about 120 m (400 ft) almost half a mile away. When Pool switched off his engine he heard a loud whine similar to an electric motor running at high speed, coming from what he now realized was no ordinary aircraft.

The object continued to dive down until it came to a halt, hovering about 2 m (6 ft) from the ground and barely 3 m (10 ft) from Pool's car. He sat staring at the object, noticing that it had a flat underside and domed topside. The object was some 6 m (20 ft) across and at one end had a sort of upward bulge, which Pool took to be a cabin. There was what seemed to be a window on one side, but he could not see through it.

After sitting in amazement for some seconds, Pool decided to get out of his car and approach the strange object. No sooner had he opened the door, however, than the disc rocked to one side, then shot off at high speed.

Throughout the 1950s, 1960s and 1970s the sightings of UFOs came largely from North America and from Europe. Critics argued that this was because the initial flying saucer stories that originated with the Kenneth Arnold sighting – see Encounter Casebook No. 1 on page 24 – had been publicized in those areas, prompting people to misreport normal objects. But gradually sightings began to be made in other areas. It later transpired that UFOs had been seen regularly over the Soviet Union and other communist countries from 1947 onwards but that the authorities there had imposed a news blackout on the subject.

Although the Soviet and communist authorities were far more able to suppress news than those in the West, news of UFO sightings did begin to leak out during the 1960s. Some of these reports dated back to the late 1940s and early 1950s, just when UFOs had begun to be seen in numbers over the USA and Europe. These reports were mostly second-hand and included few details. Only those occurring at the time the reports came out had any real interest.

Typical of the early reports is that from the Polish city of Poznan in 1957. A local newspaper reported on 31 January that an unknown aircraft shaped like a disc had been seen flying overhead, but gave no other details.

Fortunately later reports were rather more helpful.

On the warm evening of 1 July 1966 Vlasta Rosenauerova was sitting on the veranda of her house near Pilzen, Czechoslovakia, while her two grandchildren played nearby. Mrs Rosenauerova was thinking that it was time to call the children in as the sun was setting when she saw two lights approaching from the north. They were diving as they approached and had a yellowish-red colour similar to that of a candle flame. When the objects were an estimated 2 km (1.25 miles) distant one stopped, followed by the other.

The object flew northwards with an undulating motion for about five seconds, then turned abruptly and shot off to the west.

Mrs Rosenauerova described the two objects as being identical in size and shape. Each was spherical with a domed bulge on top and was about 35 m (115 ft) across. Mrs Rosenauerova pointed the lights out to her grandchildren, whereupon the younger one burst into tears. As Mrs Rosenauerova comforted her grandson the elder child, a girl of about 9 years old, called out, 'Look Grandma, it is on fire.' The left hand object was indeed spouting what seemed to be a thin column of smoke. A few seconds later both objects changed colour to bright white and accelerated away to the northeast until they were out of sight.

On 20 November 1967 the Hungarian poet Laszlo Benjamin and a friend, Peter Kuczka, were walking down Krisztina Avenue in Budapest when a large aircraft flew low overhead. Benjamin glanced up to see not the passenger aircraft he expected but a spherical object about 21 m (70 ft) across which had a thin flange or rim around its middle. The object gave out a bright white flash of light, then took off at high speed.

Three days later a Yugoslavian Communist Party official named Punisja Vuiovici was passing the shores of Lake Krupat when he saw what he at first took to be a shooting star. As he watched, however, the object came right down to hover over the water and emitted a pulsating white light rather brighter than any star. The object was conical in shape, though Vuiovici did not hazard a guess as to its size. After a few seconds it flew off.

On 21 September 1968 Nicolae Radulescu, an engineer in Romania's Ploesti oilfields, was watching a brilliant sunset from his apartment window when he saw an object flying past. It was coloured pale red or pink and shaped like a disc, though its precise shape was unclear due to it being surrounded by a mist or spray of gas. The object flew northwards with an undulating motion for about five seconds, then turned abruptly and shot off to the west.

Although the vast majority of sightings are of spherical or disc-shaped objects, not all fit this pattern. A good number of UFOs are reported to be shaped like a cigar. On 6 May 1952 farmer R. Geppart was ploughing a wheat field near Wagga Wagga in New South Wales, Australia in the cool of dawn when he saw lights approaching.

At first Geppart took little notice, taking the lights to be the headlights of a car on the road, but as

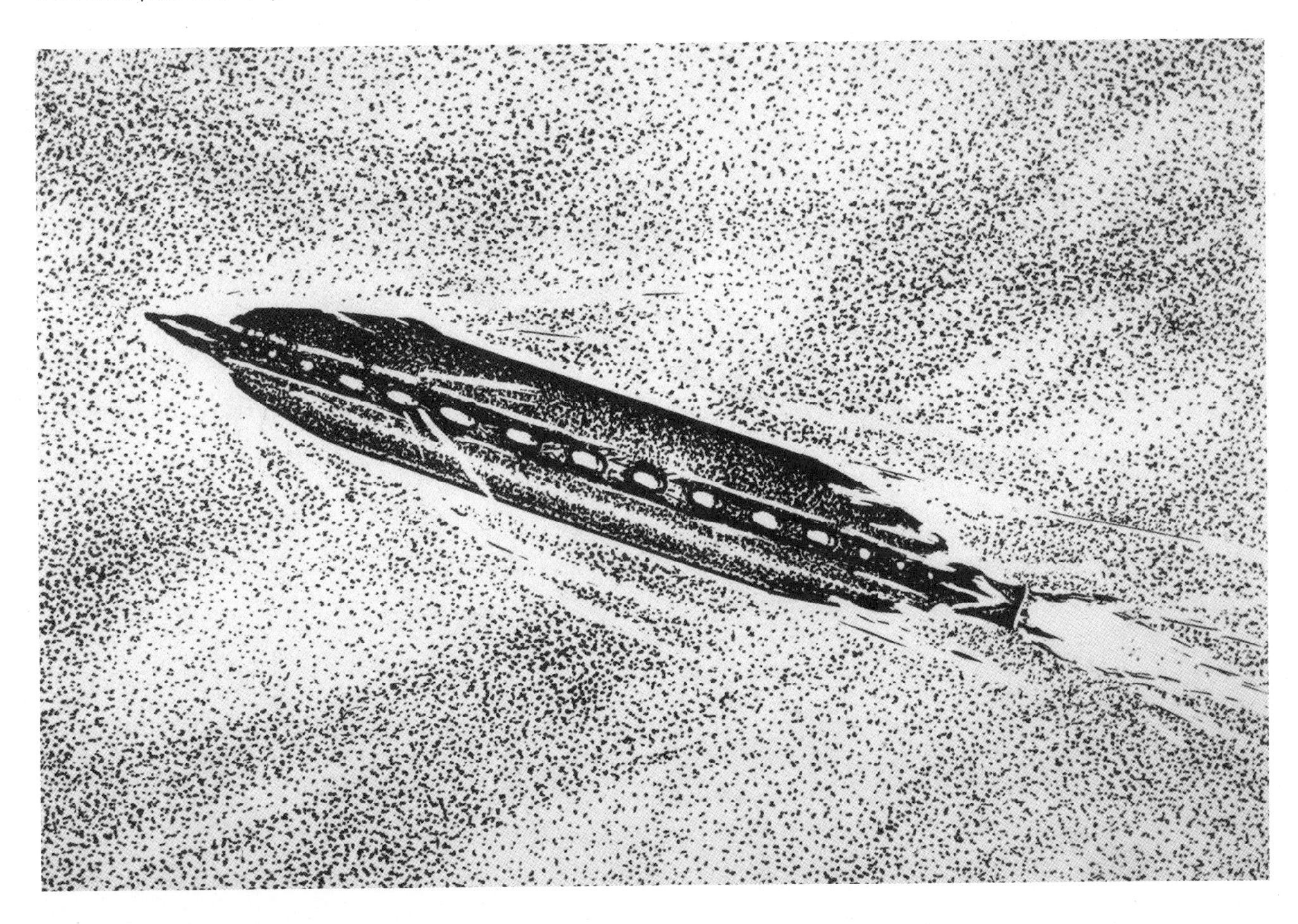

A cigar-shaped UFO with a row of lights along its length. UFOs of this shape are the most often reported after the classic disc shape.

they got closer he realized that they were in the air and not on the road itself. Now watching more closely, Geppart saw that the lights were fixed to a larger object that glowed only faintly. The object was about 30 m (100 ft) long and 6 m (20 ft) across. It was moving more slowly than an aircraft at a height of some 75 m (250 ft). It passed Geppart at a distance of just 120 m (400 ft) so he got a good look. There was a nose cone of a reddish colour and behind that what Geppart took to be round windows or portholes through which light was streaming from inside the object – these being the lights that had first attracted his attention. The object glided past in absolute silence, then picked up speed and vanished into the distance.

The silver object this time remained intact, but was followed by four orange balls as it cruised south over Cradle Hill.

Another cigar-shaped object was seen over South Australia in July 1960 by cook Mrs W. Pettifor when walking to work at a hotel in time to prepare breakfast. She saw what she at first took to be an old Zeppelin airship, though she could not imagine what such a thing was doing at Moana township. The object seemed to glow orange as if there was some bright light deep within it diffusing out through an opaque covering. The object glided through the air in serene silence for a while, then halted for a few seconds before accelerating vertically at great speed and vanishing from sight.

Very often such sightings are single events that come and go without warning. But sometimes there occur what have become known as flaps – when several sightings take place within a fairly small area over a reasonably short period of time. One of the most famous such flaps began on 1 April 1965 in the rather mundane surroundings of a council car park in Warminster, a small market town in Wiltshire, England.

Charles Hudd, an employee of Warminster Council, was reporting for work at 4.45 am when he spotted a large silver object flying through the sky. The object was rectangular in shape, rounded at both ends and moved in total silence. Hudd called the attention of three workmates to the object and all four men watched as it moved off to hover over Cop Heap, a nearby hill. As it hovered the silver object gradually turned red, then abruptly split open and turned into four smaller red balls which dropped towards the ground before streaking off to the north.

Given the date, the local newspaper editor was not inclined to believe Hudd when the report came in. Though he did not print it, he filed it for future use. That use was not long in coming. Three days later, a bank manager sighted what seemed to be the same silver object 3 km (2 miles) south of Warminster as he drove home after working late. The silver object this time remained intact, but was followed by four orange balls as it cruised south over Cradle Hill.

In the early hours of 7 April two army officers, Lieutenants R. Ashwood and P. Davies of the Welch Regiment were leaving a particularly good dinner in Warminster when they spotted a silver disc flying overhead at high speed. The object flashed over the town, swept over Cradle Hill and vanished into the distance.

The trio of sightings in one week caused a sensation in the town. The 'Thing', as the phenomenon was known locally, prompted many local people to come forward to report odd noises they had heard in the weeks leading up to the sightings. At the time they had put the noises down as being made by the army at their exercise grounds on nearby Salisbury Plain, but now people were not so sure.

At 9 pm on 3 June the Thing was back. This time the silver, cigar-shaped object was glowing brightly in the night sky as it hovered over Heytesbury, a village south of Warminster. The witnesses this time were the wife and children of the local vicar. Twelve-year-old Nigel Phillips had a small telescope, which he excitedly trained on the object. With its aid he saw a protrusion towards one end of the object which was more orange than silver. The boy drew a picture.

It was that picture, and the impeccable background of the witnesses, that catapulted the Warminster Thing into the national media. Journalists hurried to Warminster to gather stories. Some linked the Thing to Cley Hill a few miles west of the town which for centuries had been rumoured to be where the Devil sat when he gazed across Wiltshire looking for evil to do. Others made the more obvious link between the Warminster Thing and the UFOs reported elsewhere.

Reports of discs and cigar-shaped objects in the sky over Warminster continued to come in during June, July and August. Many of these seemed centred around Cradle Hill.

As Pell watched the ball gathered speed, then changed direction and came straight towards him. Pell slammed his brakes on, then swerved to avoid a collision with the great ball of fire.

On 10 August lorry driver Terry Pell of Lincolnshire was driving a load of vegetables from his home county to a storage depot at Warminster. As he came down Cop Heap Lane at around 4.30 am he was startled to see a 9 m (30 ft) diameter ball of crimson flame tumbling down the slopes of Cradle Hill. As Pell watched the ball gathered speed, then changed direction and came straight towards him. Pell slammed his brakes on, then swerved to avoid a collision with the great ball of fire (see illustration on page 45). The lorry ended up in a fence while the Thing climbed rapidly and vanished as the badly shaken Pell watched it go.

On 25 August Peter Wilsher and his fiancée Ann spotted four silver discs circling the hill. The discs were in view for several minutes, as they changed colour from silver to pale red and then flew off. The couple were on holiday from Essex and had missed the fuss about the Thing. When Wilsher mentioned the sighting in a local pub he expected not to be taken seriously, but was pounced on by locals who demanded details and then sent him to see the editor of the local newspaper.

On 27 August a public meeting was arranged in the town hall to which hundreds turned up, and hundreds more had to be turned away. Little was achieved, except that dozens of people recounted their sightings and a government official stated that the Thing was not some secret army weapon that was out of control.

Thereafter sightings of the Thing tailed off for a while. There was a brief flurry of activity around the turn of the year. On the morning of 20 December Eva Robinson saw a silver cigar-shaped object flying over the hills outside town as she walked to work. On 4 January 1966 Rosemary Bell reported a gleaming orange ball rolling through the sky. On 14 January some teenagers saw a strange triangular-shaped craft that was coloured gold pass overhead at the village of Battlesbury. Three nights later a group of children playing in fields saw a chain of white lights flash overhead.

By this time the Warminster flap had become news on a national and international scale. The focus of sightings around Cradle Hill led one group of young UFO enthusiasts to set up a nightly watch on the summit of the hill. The watch lasted for months, but not much was seen.

As with other flaps elsewhere, that at Warminster in 1965 seemed to peter out inconclusively. There gradually emerged a pattern to such flaps. Initially, there were a number of sightings of UFOs which built rapidly to a peak, with a large number of sightings being made in the space of a few days or couple of weeks. Thereafter the number of sightings declined in both number and quality as close encounters of the first kind were replaced by daylight disc or nocturnal light sightings. Finally, after a brief spate of sightings, the flap ended and quiet returned.

Some researchers have speculated that this pattern is more to do with how people react to a flap than to the flap itself. They hold that the initial large number of high-quality sightings is evidence of intense UFO activity. This is then reported in the media. In the weeks that follow people report as UFOs things that they might otherwise not have thought unusual or which could be adequately explained as normal events and objects. The final blip may either be due to real UFO activity or to media outlets reviewing the story. Thus what might in reality be only a few days of real UFO activity gets stretched out into an apparent flap lasting many months.

Whatever the truth behind flaps, dozens, perhaps hundreds, of people had definitely seen things in the skies over Warminster that neither they nor anyone else could explain. There had been nocturnal lights and daylight discs in abundance, plus some startlingly clear close encounters of the first kind. But there had been no proof and no explanations.

Those seeking either proof or explanations for the UFO mystery would have to look elsewhere.

The moment in 1965 when lorry driver Terry Pell crashed off a rural road near Warminster, England, to avoid a fiery UFO that flew low over the road in front of him.

Encounter Casebook No. 2

TYPE>>**Close encounter of the first kind** DATE>>**3 September 1965** PLACE>>**Exeter, New Hampshire, USA**
WITNESSES>>**Norman Muscarello, Patrolman Eugene Bertrand, Patrolman David Hunt, Unknown woman**

The Exeter incident, as it became known, gained fame largely because the two policemen involved refused to be fobbed off with anodyne 'official explanations' issued by Project Blue Book.

The incident began at 1 am when Patrolman Eugene Bertrand was driving along Route 108. He saw a car parked by the side of the road in a remote rural spot and pulled over to investigate. Inside the car Bertrand found a woman driver in a state of some distress. She said that her car had been followed by a bright white light in the sky that had dived down to hover over the vehicle. She had then stopped and the light flew off as Bertrand approached. After a quarter of an hour the light had not returned, so the woman drove off while Bertrand continued his patrol.

Bertrand reported back to Exeter Police Station at about 2.30 am. There he found Norman Muscarello who was shaking with fear and almost unable to talk. After a few minutes Bertrand and desk sergeant Reginald Towland got Muscarello calmed down and managed to get his story from him.

Muscarello said he had been hitchhiking on Route 150, but unable to get a lift was walking to Exeter. He had reached Carl Dining Field Farm when a group of five red lights came swooping down from the sky to hover over a house about 30 m (100 ft) from where Muscarello was standing. The red lights began to pulsate in a pattern that repeated itself. As the startled Muscarello stood watching, the lights suddenly darted towards him. Muscarello dived into a ditch, and when he peered back over the edge of his hiding place, the lights were diving down behind a line of trees as if landing in the field beyond.

> She said that her car had been followed by a bright white light in the sky that had dived down to hover over the vehicle.

Bertrand drove Muscarello back to the site of his encounter. The pair got out of the parked patrol car and looked about. There was nothing to be seen, so Bertrand radioed back that all seemed quiet. Towland suggested that the field where the lights were seen to land should be investigated. Bertrand switched on his flashlight and began advancing. He was about 15 m (50 ft) from the car when cattle at the nearby farm began to call loudly as if in alarm, then mill about in excitement. Suddenly the red lights rose up from the ground behind the trees.

A time-lapse photo enhances the stars and other aerial objects over the unlit roads of rural America. The Exeter incident took place in just this sort of landscape.

Confronted by a UFO at close quarters, police patrolman Bertrand drew his gun. Realizing that any aliens would probably have more powerful weapons than his own, he chose not to fire.

Bertrand drew his pistol and Muscarello began shouting 'Shoot it, shoot it!' But as the lights came closer Bertrand thought better of opening fire and instead raced back to the patrol car. The two men hid behind the car as the lights approached to within about 30 m (100 ft). So bright were the lights that Bertrand began to fear that he might get burnt. He scrabbled inside his car to find the radio and called for backup.

Patrolman David Hunt took the radio call at 2.55 am. By the time he arrived at Bertrand's parked car the red lights had retreated. Hunt saw them clearly enough about half a mile away. A minute or two later the lights rose higher into the sky and headed off in a southeasterly direction, accelerating rapidly as they went.

All three men were considerably shaken by the experience. When word got out they came in for a fair degree of teasing from colleagues and friends, but both Hunt and Bertrand felt that they should make a formal report. They accordingly wrote out and signed statements that they sent to Project Blue Book.

The local press took up the story, prompting the Pentagon to issue a dismissive statement that the men must have mistaken a flight of B47 military

> The cattle at the farm were clearly aware of something unusual, while both the unknown woman motorist and Muscarello were terrified.

aircraft that were flying over the area.

Bertrand and Hunt were indignant at the suggestion. Both men had spent long hours driving the lonely highways at night when aircraft of all types were flying about overhead. They felt that they knew what a B47 looked and sounded like, as well as other aircraft, and were adamant that what they had seen that night was entirely different. Moreover, Bertrand had spent some years in the USAF before joining the police and was even more accustomed to seeing aircraft under all sorts of conditions. In any case the B47 flight had passed over at around 1.30 am, and the sighting had continued until past 3 am. The two men wrote to Blue Book restating the facts and demanding that the USAF formally absolve them of making the story up or of being incompetent witnesses, a charge that might well damage their careers. After some weeks, and the sending of a formal second letter, the two policemen got the reply they wanted. The sighting had been reclassified by Blue Book as 'unidentified' and the competence of the officers accepted.

The Exeter incident is interesting not just for the details of the UFO itself but also for the reactions of those involved. The cattle at the farm were clearly aware of something unusual, while both the unknown woman motorist and Muscarello were terrified. The two policemen reacted on the night with commendable courage, especially Bertrand, but did not really know how to respond to a situation outside their experience. The USAF and Pentagon ridiculed the idea of a UFO even before they had had a chance to study the information, which prompted the townsfolk of Exeter to ridicule their policemen. Bertrand and Hunt then stuck to their story absolutely until they received a partial vindication. Thereafter both Exeter and the patrolmen dropped out of the UFO story.

It is indeed as if something utterly bizarre and inexplicable fell out of the sky over Exeter that night, then vanished.

A USAF B47 heavy bomber. The USAF at first tried to dismiss the Exeter incident as one of these aircraft seen by untrained witnesses, but the explanation collapsed when it was revealed that one witness was a former USAF airman.

CHAPTER 3

The UFOs Land

The earliest sightings of UFOs, and the majority of reports since, have been of flying objects seen mostly from a distance but sometimes from close up. Convincing as many people found these reports, they left behind them no evidence beyond the sightings themselves. Sceptics could claim that the witnesses had mistaken ordinary objects for something extraordinary, or that they might have been hallucinating or even lying.

Then sightings began to take place where the UFO left behind physical traces of its passing. In the terminology adopted by UFO researchers these are close encounters of the second kind, or CE2s. They do not happen often but when they do they can be dramatic. Studying these CE2s allows us to learn more about the UFO phenomenon.

Close encounters of the second kind and the traces they leave behind have been of enormous importance to the study of UFOs and alien encounters. Sightings alone, even when the witness accurately reports what was seen, are open to all sorts of doubts. The actual size of an observed object depends very much on how far away it was when seen, and judging distances can be notoriously difficult especially at night or when seeing an object in the air. If an object leaves physical traces, however, these can be precisely measured at leisure. Moreover the type of traces left may indicate how the object moved, how hot it was and of what material it was composed. For instance, holes cut from soil indicate a mechanical action, while holes burned in vegetation suggest radiant heat. They also discount entirely the possibility that the witness invented or hallucinated the encounter.

If an object leaves physical traces, ... these can be precisely measured at leisure.

As the type and number of close encounters of the second kind accumulated, investigators tried to make sense of the growing bank of data.

An early case to achieve widespread publicity happened on the night of 2 November 1957 at Levelland, Texas. The fact that this took place just an hour after the Russians had launched humanity's second artificial satellite probably had something to do with the publicity.

At 11 pm Patrolman A. J. Fowler answered the phone at the Levelland police station. The call was from a man named Pedro Saucedo. Immediately

The classic UFO is usually described as being a metallic, silvery disc with a pronounced flange or rim around its centre.

Fowler could tell from Saucedo's voice that he was in some distress. Saucedo reported that his truck had broken down after he saw a bright light in the sky, but that it was now working again. Unable to get many details from Saucedo and thinking the man might have been drinking, Fowler logged the call but took no action.

Some days later Saucedo and his friend, Joe Salaz, gave a rather more coherent account of what had happened. As they were driving west from Levelland on Route 116 they had seen a large flying object shaped rather like a torpedo coming towards them. As the object got closer their truck engine had coughed and then died, the headlights blinking out almost immediately after.

Saucedo got out of the truck to get a better look at the rapidly approaching object, which he thought was about 60 m (200 ft) long. It was pulsing with yellow and white light and giving off a tremendous amount of heat. The object passed by without pausing and headed off east. A few seconds later the truck's headlights came back on and Saucedo was able to start the engine. He continued his journey to Whiteface where he found a payphone from which he made his call.

> Mrs W. ... saw that the land around her car was slowly being illuminated by a whitish light.

Fowler was, meanwhile, in for a busy night. About an hour after logging Saucedo's call, he took a second call from a Mr Watkins. This caller reported seeing an object shaped like an elongated egg about 60 m (200 ft) long resting on the road a few miles east of Levelland. As he approached the strange object, Watkins' car engine had stopped and his lights had gone out. A few seconds later the object took off and headed north. The car headlights had then come on and the engine was able to restart.

Fowler had barely hung up when he got a third call from a man reporting an almost identical incident north of Levelland. A fourth call followed from a terrified truck driver northeast of Levelland. Three more calls were made the following morning, making a total of seven motor vehicles that spluttered to a stop after encountering a UFO that night.

By 1 am, Fowler had alerted all patrolling police vehicles to the bizarre reports he was receiving. Patrolmen Lee Hargrove and Floyd Gavin spotted the UFO at a distance, but it was moving too fast for them to catch it. Fire Marshal Ray Jones picked up the police radio traffic and drove out to join the UFO hunt. He too saw the UFO and although his engine promptly spluttered, it did not stop.

Once the local newspaper printed the story, over fifty local residents came forward to say that they had seen a strange light or object in the skies around Levelland that night.

A broadly similar experience to that which befell the motorists around Levelland occurred on the evening of 22 September 1974 near Launceston, Tasmania. A woman, who preferred to be known only as Mrs W. in press reports, drove to an out of town bus stop where she was due to pick up a visiting relative, parked her car by the roadside and settled down to listen to the radio.

After a few minutes the radio began to emit a high pitched whine. As Mrs W. bent forward to fiddle with the radio controls she saw that the land around her car was slowly being illuminated by a whitish light. Looking up to see the source of the light the woman saw a UFO approaching at a height of about 15 m (50 ft).

The object was about as big as a truck. The top half consisted of a curved dome that was pulsating with a bright orange-red light. The underside was shaped more like an inverted cone, though with distinct horizontal banding. The lower half was silvery-grey in colour and was emitting a pale whitish light.

Unsurprisingly the woman decided to get away as quickly as possible. She started her engine, put the car into reverse and accelerated away as the UFO reached within 30 m (100 ft) of her car. The car had gone barely 100 m (330 ft) when it ran off the road and the engine cut out. Mrs W. watched in alarm as the UFO approached, then came to a halt. It then bobbed off to its left before climbing steeply into the sky and vanishing into clouds.

As soon as the UFO was out of sight, Mrs W. leapt from the car and ran home. Her husband met her at the door and, having calmed her down enough to get the story from her, noticed that her hair, which had been permed a few days earlier, was now straight again.

Next morning Mr W. walked up the road to retrieve the car with his son. They started the vehicle without trouble, but were deeply puzzled by the fact that the front half of the car was gleaming clean, while the rear half was as dirty as it had been when Mrs W. had begun her ill-fated journey.

Somehow the UFO had not only interfered with the car radio – which never again worked properly – it had also cleaned the paintwork.

Rather more bizarre than simply switching them off was the effect on the headlights of a car travelling near Bendigo in Victoria, Australia, on the night of 4 April 1966. Ronald Sullivan was driving along a straight rural road in a remote area near Bendigo when the beams sent out by the headlights of his car suddenly began bending to the right.

Understandably puzzled, Sullivan braked to a halt intending to inspect his headlights. As he got out of his car, however, he noticed a round or domed object in a field off the road in the direction towards which the headlights were bending. The object was of indistinct shape and glowed with various, shifting colours, most noticeably red and blue. The object then rose into the air and flew off, after which the headlights returned to normal.

> **After hovering for a while, the UFO suddenly showed seven yellow lights arranged evenly along its length.**

Similarly bizarre behaviour by beams of light was reported as part of a rather more impressive UFO sighting in France in 1972. On the evening of 11 August a group of students took part in a philosophical discussion at the open air theatre in Taizé. When the main event ended about thirty-five youngsters stayed on for an informal meeting.

At about 2 am the proceedings were interrupted by a UFO about the size of a coach that came down out of the low clouds to hover over the shallow valley in front of the open air theatre. The object emitted a low but loud humming noise. It was dark grey, silhouetted against the yellow cornfield opposite, with a single white light at one end. After hovering for a while, the UFO suddenly showed seven yellow lights arranged evenly along its length. Two yellow domes then lit up on top of the UFO near to the left end. Finally five white beams of light shone down from the underside of the UFO towards the earth. Two quite separate UFOs then appeared, taking the form of orange-red balls of fire

The 1972 Taizé sighting in France was unusual for the length of time that the UFO stayed in sight – over an hour and a half – and the number of people who saw it.

that hovered to the left of the main object.

It was by this time almost an hour since the dark grey mass had first appeared. Four of the youngsters – including M. F. Tantot and Mlle Renata – decided to investigate. Picking up their torches, though they could see perfectly well by the light cast from the UFO, they climbed out of the theatre and began crossing the ploughed fields beyond to descend into the valley over which the UFO was hovering.

It took about 20 minutes to negotiate the fence and undergrowth before they could emerge on to the valley floor. Ahead of them the four could dimly make out a large, domed object about 6 m (20 ft) tall. Unsure as to whether this was a harmless haystack or other object, or something linked to the UFO, Tantot switched on his torch and shone it at the object.

The beam of his torch shone straight ahead for most of the distance, but was then deflected upwards as if by a mirror – though no mirror could be seen. The UFO then shifted its position slightly and a beam of intense white light shot out to illuminate the four youngsters beneath it. Badly dazzled, they did not see the UFO gradually lift and then shoot off at great speed, though those still in the theatre saw it go. Also gone was the domed object in the field.

The Taizé case is interesting not only for the effect that the UFO had on the torch beam, but also for the length of time for which the UFO was visible. From when it was first seen until it vanished, the UFO was in sight for over an hour and a half, during which time it hovered almost stationary while sending out various beams of light. Most UFO sightings last only seconds or a few minutes.

One sighting that had a direct effect on the witness took place at Leominster in Massachusetts, USA, on 8 March 1967. A couple were driving home just after midnight when they drove past the town cemetery. They saw a light and a trail of smoke in the cemetery and stopped to investigate. The man got out of the car intending to enter the cemetery to see if something was on fire. As he did so the bright light began to rise slowly from the ground, moving silently and smoothly.

The man put out his arm to point at the rising object and called out to his wife to look. Then several things happened at once. First the car's engine cut out, as did the headlights, radio and dashboard lights. Secondly, the man felt a mild electric shock run through his body and his outstretched arm was pulled down to slam hard on to the roof of the car. The man tried desperately to pull his arm free, but it seemed to be glued to the vehicle.

The rising object then began to emit a loud humming sound and to rock from side to side. As the object gathered speed and moved away, the car lights came back on. The man could move again, and wasted no time in starting the engine and driving away as fast as he could.

Power failures do not only affect simple circuits such as car ignition and headlight systems. On the evening of 20 October 1990, a much more dramatic electrical power system breakdown occurred in the Botosani region of Romania.

At around 10 pm that evening Virgil Atodiresei was walking home to his farm near Flaminzi. He saw a light glowing through clouds over the nearby village of Poiana, but at first took no real notice of it. Then he saw an almost spherical object descend from the clouds. It had a slightly indistinct outline, but glowed with colour as if there were several bright lights moving around within an opaque globe. Atodiresei could see it was big, but could not estimate a precise size.

Meanwhile all the lights in Poiana went out as the electrical supply failed. Power cuts in rural Romania

The strange events at Leominster, Massachusetts, began with an eerie sighting in a graveyard at night.

were nothing unusual, so most villagers went to bed. However, Professor Nicolai Bildea had his students' mathematical papers to mark. He had a lantern in an outside shed and so went out into the back yard to find it. As he left his house he noticed a flickering yellow-red light. Thinking that a neighbour's barn or hayrick might have caught fire despite the persistent drizzle, Bildea went out into the street to investigate.

> Something had been consuming electricity that night, but it was not the villagers of Poiana.

As he came round the corner of his house, Bildea saw a large object that he later described as being shaped like a tortoise shell about 50 m (160 ft) long and 12 m (40 ft) wide. Around the edge of the flat lower side he could see a number of small white lights. From four of these, beams of light were pointing down and sweeping the village. One swept over him briefly and was extremely bright. The rain had by now stopped, so Bildea ran back to get his wife and mother to see the UFO. By the time they were in the street the UFO was moving off and they got only a distant view of it. As the UFO left the rain began again, this time a heavy downpour rather than the earlier drizzle.

Half an hour or so later the electricity supply came back on. On the following day the villagers phoned the local electricity works, which sent an engineer out to check the systems. No faults or problems were found with the electricity supply which was now working perfectly. Interestingly, the engineer confirmed that the flow of electricity from the local substation had continued throughout the night, rising slightly at the time the UFO was seen. Something had been consuming electricity that night, but it was not the villagers of Poiana.

Other physical manifestations have been ascribed to UFO sightings, none more macabre than the string of animal mutilations. The apparent trail of destruction began on the evening of 7 September 1967 when Agnes King saw a fairly large, rounded object fly over fields near her home in Alamosa in Colorado, USA. At the time she thought little of it, not even wondering if the object had been a UFO or aircraft.

Next morning however, Mrs King's daughter's horse was found dead. What struck Berle Lewis, the daughter, about the corpse was the fact that the head and neck had been stripped clean of flesh while the rest of the carcass was untouched. She had seen livestock killed by dogs and mountain lions, but this pattern did not fit the way those predators ate their prey. Even more surprising was the post-mortem carried out by a local vet. The horse's brain, spinal column and intestines were missing.

The case hit the headlines and soon dozens of ranchers, farmers and pet owners were coming forward to claim that their animals had been killed and mutilated. Many of these animal killings were not linked to UFO sightings in any way and were

The head of a dead cow. The sudden deaths and apparent mutilation of several domestic livestock have been blamed on UFOs.

probably the work of local predator animals – though some of the more bizarre injuries were ascribed by police to satanists or sadists of the human variety. In other cases the UFO link was clear.

In August 1989 a flying disc was seen coming in to land on farmland near Denver, Colorado, USA. Next day two dead cattle were found in the field.

On 12 February 1992 a disc-shaped UFO with flashing blue and yellow lights around its rim was seen over farmland near Calumet in Oklahoma, USA. Next day a steer was found dead with bizarre injuries to the head and hindquarters. The wounds had not bled and analysis showed that the incisions had been cauterized as they were made, as if the implement that made the cuts had been white hot.

In March 1992 an otherwise unremarkable nocturnal light type of UFO was seen over Webster County in Missouri, USA. A cow was found lying dead near the scene of the sighting on the following day. The animal's throat had been cut and its udders removed. Again analysis showed the wounds were cauterized by great heat, but this time the blood had been drained from the carcass.

Injuries do not happen only to animals when UFOs are in the area, but also to humans. On 29 December 1980 Vickie Landrum, her grandson Colby and her employer Betty Cash had driven to New Caney, Texas, for dinner and were driving home through a quiet wooded area when Colby spotted an aircraft of some kind hovering over the trees.

The object slowly moved so that it was hovering over the road in front of Cash's car. Cash braked to a halt, thinking that an aircraft or helicopter was in trouble and was about to land on the road. The object came to within about 55 m (180 ft) of the car, then halted and began to hover.

The UFO was shaped like a diamond, hovering with one point facing downwards. It was a silvery colour and seemed to glow. Around its centre was a string of round blue lights. From the bottom of the craft there suddenly erupted a column of red flame, causing the object to rise slightly. When the flames cut out it sank slowly down, only to be boosted up again by the roaring jet. When not emitting flames, the object made a loud beeping noise.

Both Vickie Landrum and Betty Cash got out of the car to look at the object. There was a great heat emanating from it, which caused Landrum to grow uncomfortable and return to the car. After a few minutes the object flew off. When Cash tried to open the car door, the handle was so hot that it burned her fingers. She used her jacket as a glove to open the car door. The women then saw several helicopters fly low overhead as if chasing the UFO.

A flight of Apache military helicopters at dusk. The witnesses involved in several UFO sightings have reported seeing helicopters apparently chasing the strange flying objects.

They watched the UFO fly off, keeping it in sight for 5 minutes or so, then drove home.

Next morning, Betty Cash woke up suffering from what appeared to be severe sunburn and collapsed with a pounding headache and vomiting attacks. Colby and Vickie Landrum also had headaches and what seemed to be sunburn, though their injuries were less severe. By mid-morning, Betty's neck had swollen up and her skin was erupting into blisters.

All three went to see the doctor, and Betty was whisked off to hospital. She was treated for burns, but it was not until a few days later that she told hospital staff about the UFO encounter. By that time her eyes had puffed up so much that she could barely see and her hair had begun to drop out.

The condition had eased a couple of weeks later, but recurrent attacks continued for at least two years. Betty's eyesight never returned to normal, while Vickie became so sensitive to heat sources that she was unable to return to her work as a cook. The doctors were uncertain as to the precise nature of the injuries. Their best guess is that the three were exposed to intense ultraviolet radiation or to a burst of X-rays at high levels, though quite how that could have occurred at night on a remote Texas road remains unclear.

The Texans may have suffered very badly, but they are not alone in having injuries caused by UFOs. In May 1967 Canadian prospector Steve Michalak was investigating low-grade silver deposits near Falcon Lake, Ontario, when he came across an object in a forest clearing. The object was oval in shape and about the size of a large private aircraft. It was glowing with a vivid purple light.

As he approached the door, it suddenly slammed shut and shot out a stream of searingly hot vapour.

Thinking that he had found some secret military aircraft, Michalak decided to investigate. Seeing what he took to be an open door, he went to have a look inside. As he approached the door, it suddenly slammed shut and shot out a stream of searingly hot vapour. Michalak's shirt burst into flames, so he tore it off as he hurriedly ran away. Heading for the nearest town, Michalak began suffering from vomiting and headaches before he reached civilization. When he was admitted to hospital, doctors found that his stomach had been badly burnt in a neat grid-like pattern where he said he had been hit by the vapour.

Similar injuries were suffered by Finn Eero Lammi in November 1976 when he was hit by a ray sent out by a UFO that he encountered. The burns were severe, but localized and he made a swift recovery.

Rather more benign are what appear to be marks left behind by some UFOs as they pass. In July 1969 a UFO was seen hovering low over a field by the farmer's teenaged daughter and her friend. The two girls reported that the UFO was shaped like an upturned soup bowl, though with a bottom that was only slightly concave. It was coloured silver and seemed to have had a band of softly glowing orange light around its rim. The girls thought that the UFO was about the size of a truck. It made a loud whooshing noise as it climbed away from the field.

The girls were frightened by what they had seen, but the farmer refused to believe them and sent the friend home, thinking that the girls were up to some sort of prank.

The following day the farmer looked into the field where the UFO had been seen and was amazed to find that a large circular area had been blasted. The plants looked as if they had been subjected to intense heat. The leaves and stems were wilted, browned and in places dry to the touch. None of the plants had been crushed or pushed over, so whatever had affected them had not come to rest.

On 11 May 1969 Marcel Chaput, a Canadian logmill worker, was woken up at around 2 am by the sound of his dog barking furiously. Light was streaming in the window, so Chaput looked out. He saw what seemed to be an intensely bright light hovering over a field some 180 m (600 ft) away. Chaput threw on a coat and boots and hurried out with his dog.

As he emerged from the house, Chaput saw the light dim. He estimated that the bright light was about 4.5 m (15 ft) above the ground. The object then began to make a loud purring noise and rose into the air before flying off.

> **The main discs were silver in colour and the smaller domes seemed to be semi-transparent, though nothing could be seen of the inside of the objects.**

Chaput and his family went out to the field on the following day. They found three small circular marks as if dinner plates had been pushed hard into the soft ground. The marks were arranged in a triangle, in the centre of which was a rectangular mark about 5 cm (2 in) deep. Chaput guessed that the circular marks had been made by landing legs and the rectangular mark by a hatch of some kind – though he had seen none of this the night before.

On 23 November 1957 a USAF lieutenant, whose name is suppressed in the Blue Book report, was driving near Tonopah in Nevada just after dawn when his car engine suddenly cut out. Braking the car to a halt, the lieutenant got out to inspect the engine and heard a high-pitched whine over to his right. Looking round, he saw four silver disc-shaped objects sitting on the desert about 270 m (900 ft) away.

Realizing at once that he was seeing something very unusual, the officer took the time to study them. Each object was about 15 m (50 ft) across and 3 m (10 ft) tall. On top of each was a smaller dome some 1.5 m (5 ft) tall and 3 m (10 ft) across. The main discs were silver in colour and the smaller domes seemed to be semi-transparent, though nothing could be seen of the inside of the objects. Around the rim of the discs there was a dark band that seemed to be rotating slowly around the craft. Underneath the craft were three half-spheres of a dark colour which seemed to be landing gear of some kind.

Having seen all he could from the road, the officer decided to approach the objects to get a closer look. When he was about 15 m (50 ft) from the objects the whining noise increased in pitch and volume rapidly until it began to hurt the man's ears. The objects then lifted off the ground and swept off to the north about as fast as a man could run to eventually disappear behind some hills.

Once the objects were gone the officer went forward to the place where they had rested. He found that each object had left three shallow marks in the ground corresponding to the landing gear he had seen. The marks were circular, about 2.5 cm (1 in) or so deep and were arranged in a triangle about 3 m (10 ft) or so along each side.

Returning to his vehicle, the officer was able to start the motor without trouble and so drive back to his base where he reported the sighting. The report was not made public at the time and emerged only after Project Blue Book closed in 1969. A note attached to the file by Blue Book staff back in 1957 explains: 'The damage and embarrassment to the USAF would be incalculable if this officer allied

Several UFO sightings begin quite mundanely as witnesses think they are seeing car headlights shining through trees or fog. Only gradually do the witnesses realize that they are seeing something most unusual.

himself to the host of "flying saucer" writers who provide the air force with countless charges and accusations. In this instance, as matters stand, the USAF would have no effective rebuttal or evidence to disprove any unfounded charges.'

The sighting was officially explained as due to 'psychological factors', though what they were and how the unnamed officer felt about the matter was not recorded.

One of the best documented cases of a UFO leaving a physical trace that persisted after it had gone occurred in Delphos, Kansas, in 1971. At dusk on 2 November, 16-year-old Ron Johnson went out to pen his family's small flock of sheep for the night. As he did so Ron noticed a bright light in a small clump of trees about 23 m (75 ft) from where he was walking. As he watched, the light rose steadily and he could make it out to be a disc-shaped object about 3 m (10 ft) in diameter with a single cylindrical support beneath its centre. It was making a loud, uneven whine.

Ron raced back to his house and alerted his parents, who came out just in time to see the object clear the top of the trees and fly off. After waiting to see if it would return, the family nervously entered the clump of trees. They found one dead tree had been knocked over and a branch broken off a living tree, and there was a softly glowing ring on the ground.

Next morning the Johnson family alerted the local newspaper and police. The journalists got there first. They found that the dead tree had been pushed over recently and that the broken branch seemed to have been snapped off by a heavy weight pushing down from above. Although the leaves on the branch were fresh and green, the wood itself was dry and brittle, as if it had been snapped off weeks earlier.

The place where the UFO itself had apparently landed was marked by a ring of dry soil – the rest of the earth was muddy due to heavy recent rain. The surface of the dry soil was crusted and slippery. When they dug down they found that the dryness extended for more than 30 cm (1 ft) beneath the ground surface.

When the police arrived they confirmed the journalists' findings, and added another remark to their report. The trees near the landing site seemed to be discoloured compared to others in the copse.

In the weeks that followed the Johnsons noticed that the undergrowth near the landing site was growing back. Four years later it had grown back everywhere except on the ring of soil where the UFO had landed.

Rather more unusual were the marks left by a large UFO that landed in Malaysia on 6 March 2000. A group of people saw a glowing round light swoop over their village of Kampung Gobek late that night. The object seemed to hover, then dived down slowly and appeared to land in a stretch of marsh near the village. As it descended, the object sent out flashes of light as bright as, and similar to, lightning. After a few minutes the object took off again.

As it descended, the object sent out flashes of light as bright as, and similar to, lightning.

A group of more intrepid villagers went to inspect the spot where the object had landed. The vegetation had been flattened over a fairly wide area and indentations left in the soft soil. The marks were 'Y' shaped, measuring 15 m (50 ft) by 5.5 m (18 ft). Some 1.2 m (4 ft) from this mark was a second depression shaped like a crescent about 3 m (10 ft) across.

On 8 January 1981 a close encounter of the second kind took place at Trans-en-Provence, France. In itself the event was not particularly unusual in UFO terms, but it has acquired importance because it was investigated thoroughly by GEPAN, a department of the French Space Agency.

Nicolai Collini, an unemployed engineer, was busy in his back garden repairing a water pump at around 5 pm when he heard an odd whistling noise. He looked up to see a dark grey object flying towards him over a pine tree on the far side of a large open field. It was spinning as it flew down at a gentle angle. The object was circular, shaped like two plates placed rim to rim and about 1.5 m (5 ft) tall. There was a pronounced flange around the rim. On the underside were four dark patches which Collini took to be openings of some kind.

Intrigued, Collini walked over to the edge of his property to watch the object. It came down gently, coming to rest in the middle of the field about 30 m (100 ft) away from Collini. After sitting stationary for a few seconds it rose slightly, hovered briefly, then took off away from Collini to clear the trees again and leave in a north-easterly direction.

Once the object had left, Collini scrambled into the field and ran to the landing site. He found a circular indentation in the soil with, around the edge, some marks as if a sharp object had scuffed the ground. The dip in the soil was about 2.4 m (8 ft) wide and around it there was what seemed to be a ridge about 10 cm (4 in) wide.

Collini reported his sighting to the authorities and the next day men from GEPAN were on the scene. They obtained a detailed account from Collini, took measurements of the landing site, the height of the trees, the distance involved and other matters.

Using this data, GEPAN reconstructed the flight path of the UFO. The UFO was flying at around 34 kph (21 mph) when it was first seen and came to a rest 3 seconds later. This would give it a deceleration of around 0.32G. On take off it took 3 seconds to clear a line of trees, giving an acceleration of 0.45G.

French mechanic Nicolai Collini was at work on a water pump when a UFO came in to land barely 30 m (100 ft) away from him.

By measuring the indentation left in the soil by the object and the load-bearing strength of the soil, the object was estimated to have weighed around 700 kg (1,540 lb). Whatever powered a craft of this weight to the estimated deceleration and acceleration rates would have needed a thrust of around 7,171 kg (15,776 lb). The soil in the indentation was dry and brittle, indicating that it had been briefly heated to a high temperature, perhaps as much as 600 degrees Celsius.

The final conclusion of the GEPAN team was that 'These findings support the idea that the Trans-en-Provence incident represents an unconventional event.'

Perhaps the most enigmatic marks allegedly left by UFOs are the so-called 'saucer nests'. In 1966, just outside Tully in Queensland, Australia, locals found an area of marsh where the reeds had been flattened in a circle with a swirling, circular pattern. Although no UFO had been seen in the area, the local press dubbed it a saucer nest and speculated that it had been formed by a flying saucer.

Similar saucer nests turned up occasionally over the following years, but it was a group of three saucer nests that appeared in wheat fields outside Westbury in Wiltshire, England, that really sparked interest in the phenomenon. Westbury is not only close to Warminster, the site of a famous UFO flap in the 1960s, but is overlooked by an ancient white horse carved into a hillside around which swirl many legends. A UFO investigation group was quickly on the scene, as were journalists and press photographers.

Each circle was about 18 m (60 ft) in diameter. The standing wheat in the nests had been bent over close to the ground so that it lay almost flat, but the stems were neither broken nor cut. Nor was there any sign of burning. The stalks were bent down in a swirling, circular pattern, the middle of which was off-centre within the circle itself. At the edge of the

One of the more elaborate crop circles that appeared in English fields during the 1980s. The vast majority of such phenomena prove to be the work of humans.

circle there was a clear and immediate change from bent stems to those standing upright. The 'crop circles', as they were called by the British press, dominated headlines for some days, then faded from public attention.

The following year three more circles appeared, this time in a wheat field in Hampshire. Again the press were quickly on the scene and again media interest focussed on a link to UFOs, although no unidentified craft had been seen.

Weather experts came forward to suggest that a relatively rare form of fair weather whirlwind was to blame. These form when warm summer sun unevenly heats up air close to hills. The rising air forms a vortex that can spin with increasing speed to reach around 160 kph (100 mph) for short periods. Given that all the crop circles formed near hills in warm weather, the meteorologists suggested that this was the cause. The general public, however, seemed to prefer a mystery.

Thereafter, the crop circle phenomenon took on a life of its own. Circles began appearing all over southern England, particularly in the West Country, with increasing regularity. In July 1983, a complex pattern of four small and one large circle was found. Within a few years mere circles began to fall out of fashion. New designs of increasingly sophisticated and complex pattern began to appear in farmers' fields. Spirals, wheels, pictograms and assorted asymmetrical designs were found.

In 1991 two middle-aged artists, Doug Bower and Dave Chorley, went to the press claiming that they had faked the more complex designs, having been inspired by the original round crop circles. To prove their point, Doug and Dave faked a crop circle design for journalists, using rope and wooden boards to flatten the standing crops without breaking any stems and using paper-drawn patterns as templates to copy. The press then called in self-appointed crop circle experts who announced the designs genuine.

Doug and Dave faked a crop circle design for journalists, using rope and wooden boards to flatten the standing crops ...

Interest in crop circles soon waned. A few elaborate designs appeared later that year, apparently created by fraudsters other than Doug and Dave, but the media was no longer interested and by 1993 the crop circle fad was over. A few simple circles appear each year, presumably made by fair weather whirlwinds.

The crop circle sensation was unfortunate as it tended to link UFOs to frauds and hoaxes both in the media and in the public mind – at least in Britain. However it did little to explain away the more dramatic forms of evidence being left by UFOs, and nothing at all to explain the most dramatic form of UFO event, the close encounter of the third kind. And the occupants of UFOs were by this date being seen with increasing frequency.

Encounter Casebook No. 3

TYPE>>**Radar-visual nocturnal light** DATE>>**21 October 1978** PLACE>>**Bass Strait, Australia** WITNESS>>**Frederick Valentich**

The fatal encounter over Bass Strait made worldwide news reports within hours of its happening. The events were dramatic and appalling in equal measure. The sighting has gone down as a classic close encounter of the second kind.

On the evening of 21 October 1978, 20-year-old Frederick Valentich prepared to fly from Melbourne to King Island and back. His task was to collect a catch of crayfish for dinner in the officers' mess of the Victoria Air Training Corps, where he was an instructor. He had made the flight many times before, but this was the first time he had done so at night. Although young, Valentich held an unrestricted pilot's licence and was keen to increase his logged solo night flying time. The crayfish mission was an ideal opportunity to do so.

Valentich took off in a Cessna 182 at 6.19 pm and at 7 pm passed over Cape Otway lighthouse. He was now flying over Bass Strait, the stretch of sea between Tasmania and mainland Australia.

At 7.06 pm Valentich radioed Melbourne Flight Control to ask if there were any other aircraft in his area. Melbourne replied that no known aircraft were around. So far as they were concerned Valentich was alone in the night sky.

There was a slight pause, then Valentich reported that a large aircraft showing four bright lights had just flown by about 300 m (1,000 ft) above him. This made the incident a near miss in air traffic control terms. The staff at Melbourne asked Valentich to confirm that he was reporting the close presence of a large aircraft. Valentich did so and repeated that the object had passed him at speed.

Valentich then asked Melbourne to check with the Royal Australian Air Force (RAAF) to see if they had any aircraft in the area that had not been advised to civilian air traffic control. Such things did happen, and it was clear by now that Valentich was becoming concerned. Melbourne promised to do so.

At 7.09 pm Valentich was back on the radio. 'It seems to be playing some sort of game with me,' he reported. Melbourne asked him if he was still unable to identify the aircraft. 'It's not an aircraft,' came the surprising response, then the harsh crackle of static cut in, making the rest of Valentich's words inaudible. When his voice came back he was saying, 'It is flying past. It has a long shape. Cannot identify more than that.' There was a pause, then Valentich blurted out, 'It's coming for me right now,' followed by another pause. Then Valentich seemed to calm down. 'I'm orbiting and the thing is orbiting on top of me. It has a green light and a sort of metallic light on the outside.'

At this point Melbourne received a response from the RAAF confirming that no military aircraft were

The lighthouse at Cape Otway, Victoria, overlooking the wild southern seas over which Frederick Valentich's aircraft vanished after a close encounter with a UFO.

over Bass Strait. Valentich replied by saying that the strange object had vanished. All seemed well until at 7.12 pm Valentich suddenly came back on the radio. 'Engine is rough and coughing,' he reported. Then a few seconds later he said: 'Unknown aircraft is now on top of me.' There was then another burst of static. Then silence.

Melbourne repeatedly tried to contact Valentich but there was no answer. King Island airfield staff were alerted to watch for the approaching Cessna. Neither sight nor sound of the aircraft was detected. At 7.28 pm Melbourne ordered that a search should begin.

Valentich had with him in the Cessna a standard life jacket and radio beacon, which was activated automatically if it fell into water. No radio signals were picked up from it. At dawn on the following day the RAAF began an exhaustive search of the Bass Strait by air. A small oil slick was seen near King Island, but that could have come from any one of a dozen sources. No sign of Valentich or his Cessna was ever found.

In the following weeks many attempts were made to explain what had happened. One suggestion was that Valentich had somehow turned the aircraft upside down and was in fact seeing the reflection of his own lights in the sea, not a UFO above him. However, he had the strange aircraft in sight for about 7 minutes, and the Cessna 182 can fly upside down for only 30 seconds before the fuel system collapses.

It seems likely that Valentich saw something extraordinary over the Bass Strait that night. Whether the UFO directly caused his death, or disappearance, there is no way of knowing. Perhaps Valentich made some catastrophic error while concentrating on the UFO instead of his instruments. Or perhaps the UFO did have a hostile intent.

Encounter Casebook No. 4

TYPE>>**Close encounter of the second kind** DATE>>**9 November 1979** PLACE>>**Livingston, Scotland** WITNESS>>**Robert Taylor**

When Scottish forestry worker Robert Taylor walked into a clearing in a wood outside Livingston, west of Edinburgh, he expected to see nothing more exciting than his pet dog chasing rabbits. What he really did see changed his life.

It was 10 am when Taylor left his pick-up truck to walk through the dense stands of conifers. His task was to check the fences and gates surrounding the forest and ensure that no sheep from neighbouring farmland had got in. His dog, an Irish setter named Lara, had run ahead of Taylor but was nowhere to be seen. Presumably it was somewhere in the forest searching for rabbits.

In the centre of the clearing was a dark grey object about 6 m (20 ft) across. It was round with a high-domed top and a narrow rim projecting out from its base. Standing up from the rim were a number of poles topped by what looked like small propellers. At intervals around the base of the dome were darker round patches that were almost black in colour. The object seemed to be hovering slightly above the ground, but was emitting no noise. It was, however, giving off a strong smell akin to burning rubber.

Understandably, Taylor came to a sudden halt as he stared in some surprise and no little alarm at the object. Almost at once, he realized that he was being approached by two small round black balls that came from the direction of the grey object. Each ball was a little under a metre in diameter and had half a dozen straight legs sticking out from it. The balls rolled towards him on the legs, making a soft sucking noise as each leg touched the ground.

After coming across the craft, Taylor was approached and attacked by two black spheres with spiked legs.

Things were happening fast. Before he could back off the two spheres had reached him and each pushed out a leg to grab hold of his leg with another soft sucking sound. The balls began to move back towards the object, dragging Taylor with them. Now alarmed, Taylor struggled to get free. The burning stench increased in intensity to the point where Taylor found he could barely breathe. Gasping for air and trying to fight off the black spheres, Taylor felt himself growing dizzy and losing consciousness.

The next thing he knew was when he woke up some 20 minutes later lying face down on the grass, with Lara moving about agitatedly nearby and whimpering. The strange objects had all gone. Taylor's trousers were torn where the spheres had grabbed him. One of his legs was badly bruised and his chin was cut and bleeding. He tried to stand, but his legs were weak so he began to crawl back towards his truck. When he tried to talk to Lara he found he could not utter a sound. Reaching the truck, Taylor tried to radio his base, but was still unable to speak. He headed home, his house being closer than the forestry base.

When he staggered through the door, Taylor was at last able to talk. He gasped out his tale to his wife, who phoned Taylor's boss Malcolm Drummond who in turn called a local doctor, Gordon Adams. Drummond arrived first with a team of workmen and headed into the forest to investigate.

When the men reached the clearing they could not at first see anything. Then one of them spotted some strange marks in the grass. Drummond ordered the men out of the clearing so that they did not disturb the ground and called in the police.

Dr Adams had meanwhile arrived and examined Taylor. All the tests came up normal, though Taylor by this time had a pounding headache to add to his minor bruises and cuts. Adams suggested an X-ray, but the local hospital was too busy and by the time the X-ray machine was available the headache had gone so Taylor never bothered having the test.

Gasping for air and trying to fight off the black spheres, Taylor felt himself growing dizzy and losing consciousness.

The police took away Taylor's clothing for forensic tests while other officers studied the marks in the clearing. There were two different sorts of marks to be found. The first consisted of two parallel tracks some 2.5 m (8 ft) long and 30 cm (1 ft) wide. These were formed of crushed grass as if an enormously heavy weight had rested on them. Around these tracks were two circles of holes driven into the soil. Each hole was circular, about 10 cm (4 in) across and 15 cm (6 in) deep. There were 40 holes in all, each of them driven down at an angle away from the tracks.

Checks with forestry workers revealed that no heavy machinery had been used in the clearing for months. The marks must have been left by the object encountered by Taylor. Police forensic checks showed nothing unusual about the soil samples taken from the holes or beneath the tracks. Taylor's clothing likewise had nothing unusual about it. The tears to the trousers were consistent with them being tugged violently by blunt hooks rather than being cut.

Everything the police found was entirely consistent with Taylor's story. But what the object was and why it was in the forest the police could not explain at all.

CHAPTER 4

The Aliens Emerge

While UFO reports had been piling up for some time in the late 1940s and early 1950s, speculation was rife as to what the objects were. Some thought that they might be secret weapons being tested by the military of the USA or Soviet Union. Others thought that they might be natural phenomena of some kind. Increasingly popular was the idea that the UFOs were alien spacecraft. This notion began to gain some support when evidence started to be reported that the UFOs were solid, mechanical objects and that they were crewed by intelligent beings who were very definitely not human.

Early on a bright October morning in 1965, Bill Hertzke was out riding on the Circle Jay Ranch some miles northwest of Calgary, Canada. He was a working cowboy given the task of bringing cattle down from the hills to the low ground for the winter. As he drove a small herd of cattle along a minor valley his horse suddenly shied. Looking round to see what had caused the problem, Hertzke saw what at first looked like a crashed aircraft. He rode up to investigate.

The object turned out to be a silver-grey craft about 5 m (16 ft) long and 3.7 m (12 ft) across. The sides had flanges or small wings shaped like those of a stooping bird of prey. On top of the front of the craft was a dome made of glass or some other transparent material. There did not seem to be any obvious joins between the dome and the craft, nor were there any panel edges or any markings on the object itself. The craft seemed to have a single seamless skin of material. Inside the dome were two seats, each big enough to hold a child. There was a door about 0.9 m (3 ft) tall set into one side of the craft. It was slightly open as if the two crew members had just popped out for some reason and would soon return. The craft was making a faint humming sound.

> **It was slightly open as if the two crew members had just popped out for some reason and would soon return.**

Hertzke was by now intrigued. He had a good look around the area for the crew, but could find nobody so he returned to the craft and began to study it more closely. In front of the seats was a flat screen about 35 cm (14 in) across, though nothing was displayed on it. Along the top of the screen was a row of five dials with markings on them which did not look like any sort of writing with which Hertzke was familiar. Below the screen were

Canadian cowboy Bill Hertzke was rounding up cattle in the early morning when he had an unnerving and significant encounter with a landed UFO.

more dials and some switches or knobs. Behind the seats was a wall with a closed door in it. Beside the door were what might have been lights. Hertzke looked to see if there were any jet exhausts or rocket outlets, but there were no holes or apertures other than the door.

At this point, Hertzke became suddenly unnerved, as if he were being watched. 'I was scared,' he later reported. 'I figured that if I didn't bother them, they wouldn't bother me.' So he left the craft and drove the cattle down the trail. When he got home a few days later it was to learn that sightings of an orange-coloured UFO had been made nearby a day or two before his encounter.

The Hertzke sighting makes a useful link between close encounters of the second kind and close encounters of the third kind. Strictly speaking this should be classed as a close encounter of the second kind. The object was seen at close quarters and had a direct physical impact on its surroundings, flattening the vegetation on which it rested. It goes beyond most such sightings, however, in that Hertzke was able to study the object at close quarters for some time. What he described was quite clearly a machine of a very solid nature – albeit one of highly advanced design that he did not recognize.

The humanoid was rigid and hard, while the skin had the texture of charred wood.

It is the observation of doors, seats and controls that puts this in a quite different category from most close encounters of the second kind. Quite clearly this UFO did have a crew, though Hertzke did not see them. Given the size of the doors and shape of the seats the ufonauts must have been humanoid in shape, just under 0.9 m (3 ft) tall and slightly built.

Rather more sensational was the sighting by Dr Enrique Botta in 1950, though he did not talk publicly about the event until 1955 when persuaded to do so by friends. Botta was a former pilot aged about 40 who, in 1950, was employed as an engineer, working on a construction project in the rural area of Bahia Blanca some 120 km (75 miles) from Caracas.

Botta was driving back to his hotel one evening when he saw a strange object resting in a field. He stopped the car to have a better look. The object was shaped like a domed disc made of a silvery metal. It had no legs or landing gear and seemed to be resting slightly askew. There was an open door on one side.

Getting out of his car, Botta walked over the field toward the object and peered through the door. Inside he could see a small, empty room lit by a vague glow and a flashing red light. As Botta touched the object he noticed that although it looked as if it were made of metal, the skin of the craft had a jelly-like softness. Walking inside, Botta passed to a second, much larger room.

In that room Botta saw a curved bench or sofa on which sat three figures facing away from him. Each figure was about 1.2 m (4 ft) tall and dressed in a tight-fitting overall that reached to the neck. The heads were rather large and looked bald.

Botta stopped in alarm, but when the figures took no notice of him he approached them. As he got close he saw that the three figures were facing what he assumed to be a control panel. It was filled with gauges, lights and what seemed to be meters. Above the panel floated a transparent sphere which rotated slowly.

As the figures still took no notice of him, Botta reached out and touched one. The humanoid was rigid and hard, while the skin had the texture of charred wood. Believing that the beings were dead,

Engineer Enrique Botta entered a grounded UFO in 1950 to find three crew members, apparently dead and slumped over a control panel. The craft had vanished by the time Botta returned with colleagues to take possession of it.

In 1952 Kansas radio worker William Squyres encountered a UFO inside which he could see a pilot. This was one of the very earliest close encounters of the third kind.

Botta fled. He dashed to his car and drove off to the hotel where he and fellow his engineers working on the project were staying.

Botta blurted out his story to his two closest colleagues. One of these men had a gun he used for hunting, and suggested that all three men should return to the craft to inspect it further. It was by now dark, so the three men decided to go in the morning.

Next day the three men drove back to the site of the encounter, but the disc-shaped craft had gone. All that was left was a small pile of ashes. One of Botta's friends stooped to touch it, but found it was hot and his hand turned purple, so he dropped the ash. Botta, meanwhile, had spotted a UFO circling high overhead. It was shaped like a cigar and pulsed with a red glow. After a few minutes it flew off and the men were left alone.

The encounter was not yet over. Later that day Botta collapsed with a fever and was rushed to hospital. His skin came up in a rash and began to blister. A test showed no sign of radiation and the doctors thought that a very severe case of sunburn was the most likely explanation, although Botta had not been out in the sun much due to his job. Botta recovered after a few days and returned to work. He decided not to talk about the incident, but later as news of UFO sightings became more common in South America, he elected to speak out.

Intriguing as the Botta and Hertzke encounters were, they gave no firm clues as to the behaviour of the ufonauts and why they were visiting the areas that they did. Nor was there any indication as to what the ufonauts were or where they came from.

What was needed, researchers thought, was an encounter with a living ufonaut – in other words a close encounter of the third kind. In fact such sightings began to be reported before the Hertzke encounter took place and before Botta made his encounter public.

Perhaps the first case of a close encounter of the third kind to be made at the time it happened, rather than in retrospect, occurred on 25 August 1952. William Squyres, a radio worker in Pittsburg, Kansas, was driving to work through farmland at 5.30 am along Highway 60. As he drove he saw something odd in a pasture field that usually held cattle. As he drew closer he saw the object was silvery-grey in colour, about 21 m (70 ft) across and 3.7 m (12 ft) high. It was shaped, Squyres said, rather like two soup bowls placed rim to rim and then flattened somewhat.

Squyres drove up until he was alongside the field and only about 90 m (300 ft) from the object. He then stopped his car and studied the object more closely. He could now see that the underside of the object had a faint bluish glow to it. Around the edge of the object was a rim or walkway. From this rose a number of vertical poles, each topped by what seemed to be a spinning propeller. The object was making a dull, throbbing noise.

At one end of the object there were what seemed to be slightly opaque windows through which indistinct objects could be seen moving. At the other end was a completely clear window through which a man could be seen apparently fiddling with controls or instruments. So far as Squyres could make out, the man was entirely human.

Fascinated, Squyres got out of his car to take a better look. Instantly the object began to rise vertically, the throbbing noise increasing in volume as it did so. The object continued to rise slowly for some seconds, then suddenly accelerated and flew off at high speed.

So far as Squyres could make out, the man was entirely human.

Squyres reported the incident to the USAF in case he had seen some sort of advanced foreign aircraft. An air force officer turned up a couple of days later and asked to be shown the site. He found that in the centre of the field the long grass had been squashed flat in a circle about 18 m (60 ft) in diameter. The grass stems had all been bent over but not broken, and formed a swirling spiral pattern. The officer collected samples of grass and soil and sent them off for analysis – which later found nothing at all unusual.

The sighting of the UFO was in many ways routine and shows many of the usual features of a close encounter of the second kind. The sighting of a pilot for the UFO inclined the Blue Book investigator to put the incident down as a

When a bright light shining into her room awoke Mary Starr in 1957, little did she realize that she was about to have a close encounter of the third kind.

hallucination, if it had not been for the flattened grass. In the event it was put down as unexplained.

Not too different was the sighting at Old-Saybrook in Connecticut, USA. On 16 December 1957, Mrs Mary Starr, a retired teacher, was awoken at about 2 am by a bright light shining into her bedroom. She looked out of the window to see a cigar-shaped object hovering about 4.5 m (15 ft) away in her backyard.

The object was about 9 m (30 ft) long and dark grey in colour. Along the side was a series of square windows from which a bright light streamed out. Through the windows, Starr could see two men walking about. The object was only about 1.5 m (5 ft) tall, so Starr estimated the men to be about 0.9 m (3 ft) tall, but said that they were otherwise quite human. They seemed to be wearing jackets that flared at the waist and square helmets that carried a reddish-orange light on top.

Mrs Starr opened her window and leaned out to get a better look. At this point a third figure came into sight through the windows. Then the light inside was switched off and the outer skin of the object began to glow with a dull blue colour. What looked like a wire antenna then rose out of the top of the object and began to emit sparks.

After about 5 minutes the antenna slid back into the object, which began to glow more brightly. The UFO then moved off as a row of small, circular lights appeared around its centreline. After clearing the yard fence, the object tilted upwards and accelerated out of sight.

Another similar sighting was made barely a week later by Mrs Suzanne Knight at Seat Pleasant, Maryland, USA. At about 9.30 pm the newly married Knight was at home with her sister, her husband being absent on business. She was tidying up the kitchen when she heard a loud buzzing noise coming from the backyard. Looking out, she saw what she at first thought was an aircraft diving down to crash into a field.

The object was shaped like an aircraft fuselage, though without wings or a tail, and was plunging down at a steep angle and at high speed. As Mrs Knight watched in alarm, the object pulled out of its dive and came to a standstill, hovering at a height of about 90 m (300 ft) some 75 m (250 ft) from the house.

The sides of the object had what looked like square windows through which streamed a bright yellow light. Through the windows Mrs Knight could see several square objects that she later likened to filing cabinets with sloping tops. On top of the object was a dull red light or glow that seemed to be rather diffuse. Underneath the object was slung a long yellow tube or box which also had windows.

At the front of the object was a larger window through which Mrs Knight could see a man. He was sitting bolt upright and staring straight ahead without moving. He had on a hat or helmet of some kind. His skin and clothing appeared yellow, though Mrs Knight thought that this might be due to the yellow light. Otherwise he looked quite human.

After watching the object hover for what she thought was a minute or more, Mrs Knight left the window to phone the local newspaper and call her sister. The newspaper phone was engaged, and her sister did not reply to her shouts.

Mrs Knight then returned to the kitchen window in time to see the object climbing back into the sky. The internal lights had gone out and the object was now more reddish in colour than silver. It began to shimmer as if it were being seen through a heat haze, then it accelerated away.

Mrs Knight's sister arrived at this point to ask why she had been calling for her. Mrs Knight told her of the UFO, but the sister merely laughed and said that she must have been dreaming. Her husband was more sympathetic and suggested she report the matter, which she did.

James Moreland, meanwhile, was perhaps concerned rather than sympathetic when his wife reported a sighting in 1959. At the time, he was employed by the Royal New Zealand Air Force. Realizing that what his wife had seen was most unusual he insisted that they report the matter to the local police, sparking off a diligent and wide-ranging investigation.

The Morelands had a smallholding near the town of Blenheim, New Zealand, on which they grew some crops and kept cattle among other livestock. On 13 July, Mrs Moreland got up at her usual time of 6 am and went out to milk the cows before turning them out to pasture. The cow barn stood a couple of hundred metres away from the house on the far side of a paddock.

Suddenly two bright green lights burst from the clouds and dived down towards the Moreland farm.

As Mrs Moreland walked over the field she noticed that a greenish glow was illuminating the low clouds as if an aircraft were shining a searchlight on to them from above. Suddenly two bright green lights burst from the clouds and dived down towards the Moreland farm. Mrs Moreland watched them approach, and began to think that there was something decidedly unpleasant about the shade of green. 'It was a horrid sort of colour,' she reported later. 'My first thought was, "I shouldn't be here," and I made a dive for the trees on the other side of the paddock.'

A New Zealand rural farmstead. It was at a farm just like this that Mrs Moreland had her classic encounter in the early morning of 13 July 1959.

From her shelter, Mrs Moreland watched the lights get closer, then realized that they were in fact attached to the underside of a large, dark grey object that she had not seen in the dawn gloom. The object was round, about 9 m (30 ft) across and perhaps 3 m (10 ft) tall. It came to a halt, hovering some 4.5 m (15 ft) above the ground and 12 m (40 ft) from Mrs Moreland. As it hovered, a series of yellowish-red spurts of fire burst out from its rim. Mrs Moreland took these to be jets keeping the craft in position. There was a constant hum the whole while and the object seemed to be radiating heat.

Mrs Moreland then saw a glow on top of the craft. This was a light coming from inside what appeared to be a round, transparent dome. Inside the dome were two humanoid figures. The figures seemed to stand about 1.8 m (6 ft) tall and were dressed in close-fitting suits of a silver metallic fabric. Both men wore helmets that rose like cylinders from their shoulders to a flat top.

One of the men stood up and leaned forward as if staring out of the dome to study the farm and barn.

He then sat down again and the jets grew brighter and the hum increased in pitch. The object tilted to one side, then climbed up to disappear back into the clouds. It left behind a noxious smell rather like burnt pepper.

Mrs Moreland was deeply unnerved by what she had seen. She waited in the trees for some time, but when the object did not reappear she gingerly went to the barn and began milking the cows. She soon stopped, however, and went to wake her husband and tell him what she had seen.

The police were called at 7 am and were soon on the scene and taking the incident seriously. They found no marks or traces on the ground, but investigations soon turned up another witness. A man in Blenheim had seen the green object diving down from the clouds, then climbing again. He had not wanted to report the sighting, but did so when he heard of Mrs Moreland's encounter.

The UFO and its crew that visited Idaho Falls in Idaho, USA, was likewise seen by more than one witness, though only one saw the crew. On 8 December 1967, Marilyn Wilding, aged 15, was at home looking after her younger sister and brothers while her parents were out. At about 7.40 pm, she stepped out on to the front porch as a friend of hers was due to come over and keep her company.

Marilyn was instantly aware of a light coming from above her. Looking up, she saw a white object about the size of a car hovering about 15 m (50 ft) above the house. The object then tipped and rotated so that Marilyn could see the upper surface, which carried a transparent dome. Inside the dome were two figures who were moving about, but the light was so bright that she could see few details.

The object then turned orange in colour and began to spin more quickly. It began to move off to the north, at which point Marilyn called for her siblings. They came out in time to see the disc-shaped UFO flying off. Neighbours later confirmed that they had seen a bright light from the direction of the Wilding house, but had paid no attention to it.

All these reports of humanoids have them inside their craft. They may look out, but seem largely indifferent to their surroundings. But other reports soon began to emerge that had the ufonauts outside their craft and very definitely doing something.

One early report came from France. On 10 September 1954 Marius Dewilde saw a UFO hovering over a railway line near Quarouble, close to the Belgian border. On the ground underneath the

Some UFOs seem to be particularly drawn to railway lines and other sources of high voltage electricity.

object Dewilde saw two figures about 1.2 m (4 ft) tall who looked like men in old-fashioned diving outfits. The object and figures soon vanished, leaving the railway tracks warped, apparently by extreme heat.

Three weeks later a fairly similar report was made by Georges Gatay, a building site foreman working at Marcilly-sur-Vienne. He spotted a UFO shaped like an upturned bowl hovering 90 m (300 ft) or so from his work site. A human-like figure standing about 0.9 m (3 ft) tall and wearing a helmet of some kind was standing nearby. Gatay went to investigate, but the figure vanished and the UFO climbed rapidly out of sight. Gatay and some of his men suffered from insomnia and loss of appetite for some days afterwards, but then made a full recovery.

On 21 October 1954 Mrs Jessie Roestenberg was preparing her two children for bed at her home near Ranton, Staffordshire, England, when she became aware of movement outside the house. Looking out, she saw a disc-shaped UFO hovering about 15 m (50 ft) away. On top of the UFO was a transparent dome inside which could be seen two humanoids. The two figures were of about normal human size and dressed in tight blue ski-suits. Both had blonde hair and seemed to be studying the Roestenberg house. When they saw the three humans looking out of the window, the ufonauts seemed fascinated, coming closer to get a better look.

Mrs Roestenberg promptly pushed her frightened children under a table and awaited developments. After a while, nothing had happened, so Mrs Roestenberg risked a glance out of the window. The UFO and its occupants had gone.

Another mother and child to be frightened by a UFO and its occupants was Mrs McMullen and her daughter Sharon, though on this occasion Mr McMullen and two of the daughter's friends from school were also present.

The encounter at Lakeland in Florida, USA, began early on the evening of 15 October 1967 when the family television began to malfunction, with the picture breaking up and the sound being interrupted by bursts of static. As the family tried to work out what was causing the interference the smell of ammonia began to seep into the house.

When they saw the three humans looking out of the window, the ufonauts seemed fascinated ...

At this point one of the students glanced out of the window and saw an object flying towards the house from a nearby hill. She called the others over to watch. The object was round, about 4 m (13 ft) in diameter and 2.4 m (8 ft) tall. On top of the disc was a transparent dome. As the UFO got closer the McMullens and their friends were able to make out two humanoid figures inside the dome. The figures were dressed in tight white clothing and had some sort of helmet on. The UFO stopped when it was about 45 m (150 ft) from the house and hovered while the occupants seemed to be studying the house. It then drifted off to one side, accelerating sharply and vanishing from sight.

The object had been in view for about 20 minutes in all. Soon after it left, the smell dissipated and the television returned to normal.

A broadly similar UFO was seen near Hanbury in Staffordshire, England, on 20 November 1968. The Milakovic family were driving towards the village when they saw several rabbits bolt out of roadside vegetation and across the road. Looking to where the rabbits had come from, Mrs Milakovic saw a brightly shining object in the field.

The object had a dark underside shaped like a wide bowl with a transparent dome placed on top. From within the dome came varying white, orange and green lights. Several humanoid figures could be seen moving about inside.

Mr Milakovic braked to a halt and the two parents climbed out of the car to get a better look at the object. The UFO then took off and flew over the road to hover on the other side of the car. As it passed, the Milakovics felt a wave of warmth as if the UFO was extremely hot and radiating heat. The UFO then began to wobble from side to side and to move in a series of jerky, random jumps. Worried that whatever the object might be it was going out of control, the Milakovics got back into their car and drove off. The last Mrs Milakovic saw as she glanced back was that the light had grown more intense and the UFO seemed to be climbing.

Many sightings of UFOs and their crews seem to indicate that they are taking an interest in water. In November 1971 Maltese fisherman Pawlu Zammit and his boat's crew were pulling nets into their boat some 32 km (20 miles) south of Malta when they

Several ufonauts were visible inside the transparent dome of the mysterious craft seen by the Milakovic family beside a rural road in England in 1968.

In 1971 a crew of Maltese fishermen came across a UFO resting on the surface of the Mediterranean. It emitted a bright beam of light, and they saw several small figures walking on the surface of the craft.

suddenly became aware of a large dark object floating on the water about 270 m (900 ft) away. The fishermen had not seen it arrive, but thought that it was most likely to be a military submarine, though they had never seen nor heard of one quite like this.

A brilliant light then flashed out from the object, making the fishermen look away. When they looked back they saw a number of small men, each about the height of a 10-year-old boy but otherwise apparently human, walking about on top of the object. Several of the men seemed to be wearing belts to which were attached tools or pieces of equipment of some kind. After a few minutes the humanoids jumped through an opening in the object's hull and disappeared.

The bright light then returned, causing the fishermen to look away once more. When they looked back the object was gone. The sighting was reported and, along with other similar events, was put by some UFO researchers into the new category of 'unidentified submarine object', or USO. Many felt, however, that this was so similar to other UFO sightings that it should be classed simply as a UFO that had landed on water rather than on land.

Also associated with the sea were the UFOs spotted off the coast of the Dominican Republic near the village of Boca Chica on 24 June 1977. A Mr A. S. Cruz was driving along the coast road when he saw a light out to sea. When the light began to move faster than a boat should, Cruz stopped to look. The light came closer and as it approached Cruz could see that it was a disc-shaped object that was glowing white and flying some 6 m (20 ft) above the water.

The object came to a halt and let down a tube that reached the sea surface. Cruz assumed that it was sucking up water. He then noticed that two windows had appeared near the edge of the UFO. Through each there was peering a face that, although human in overall shape, was distinctly odd in a vague indefinable sort of way.

The UFO lifted its tube and began moving towards the shore. Cruz got back into his car intending to drive off, but it would not start. Another UFO similar to the first now came into view and both moved to hover over Cruz's car. After a few terrifying moments, the UFOs flew off and soon after that Cruz was able to start his car.

A broadly similar account came from Guaiuba in Brazil. On the evening of 23 March 1978, retired factory worker Joao Ribeiro was strolling along the seafront when a bright light caught his attention. The light flew in low from the sea, swooping down to a height of about 9 m (30 ft), around 30 m (100 ft) away from where the astonished Ribeiro was watching.

The light could now be seen to be a disc-shaped UFO about 7.5 m (25 ft) across. It appeared to be made of a silver metal and had flashing blue, yellow, green and purple lights winking around its rim. On top of the UFO was a silver dome.

As Ribeiro watched, two humanoid figures appeared from the dome and walked over the upper surface of the UFO. The figures were about 0.9 m (3 ft) tall and appeared to be slightly built. They had large, bald, rounded heads with staring eyes and enormous pointed ears that flapped about. One carried a box, the other a long tube.

As Ribeiro watched, two humanoid figures appeared from the dome and walked over the upper surface of the UFO.

The figures walked to the edge of the craft and lowered the tube over the side so that it dangled down into the water. They seemed to be pumping water up into the box, though how Ribeiro could make this out remains unclear.

After about 10 minutes the humanoids re-entered the UFO through the dome. The UFO then dipped down until it almost touched the water, stirring up the calm surface as if a sudden gust of wind had ruffled the sea. Then it flew off at high speed.

But it is not only sea water that has its attractions – fresh water is also of apparent use to UFOs. On 22 March 2001 Vinicius Da Silva and Marta

Rosenthal were driving along the banks of the River Tocantis in Brazil after a day's fishing. Da Silva, who was driving, felt a sudden bump as if they had driven over an object lying in the road and braked to a halt. He got out to see what he had hit, and if it had done any damage to the car.

As he was peering at the road behind them, Da Silva heard Rosenthal scream. Looking round, he saw her pointing out towards the river. Hovering over the water was a disc-shaped object with a flat upper side on which was standing a short humanoid figure about 1.2 m (4 ft) tall. The object was a silvery metallic colour with a dull lustre to it.

The humanoid was standing on the edge of the dish dangling what looked like a hose into the river. After a few moments it pulled up the hose and climbed into the UFO by way of a small hatch. The UFO then began to grow brighter and shinier and slowly climbed into the sky. It then accelerated sharply and shot out of view.

Several UFOs are seen to hover over water, sometimes lowering a pipe apparently to pump up water for some mysterious purpose.

The humanoid was standing on the edge of the dish dangling what looked like a hose into the river.

What most of these sightings have in common is that the UFO is seen low over the water, or apparently resting on it, while the crew potter about outside the craft on various duties. The most significant of these seems to be the placing of a tube into the water. The tube is usually described as being not part of the UFO itself but attached to a separate box or tube mechanism.

Interestingly the witnesses almost invariably state that the UFO sighted was sucking water up from the sea or river, but there seems to be no evidence to support this. From the descriptions given, the ufonauts may just as well have been pumping something into the river or sea. Whether the UFO needs water to function correctly or if the aliens need to drink water is unclear. They might be discharging waste material or even pumping some sort of chemical into the body of water for reasons of their own.

While some UFO crews seem obsessed with water, others take an interest in flora and fauna. The behaviour of the short aliens seen by Jose Alves of Pontal, a remote Brazilian town 240 km (150 miles) northwest of Sao Paulo, in November 1960 was typical.

Alves was fishing in the Pardo River when he saw an aircraft coming towards him. As he watched, it soon became clear that this was no ordinary aircraft. It flew with a strange wobbling motion and had no wings, tail or engines. The object came to rest close to Alves and was about 4.5 m (15 ft) across.

Suddenly a small door opened and out sprang three men. Each man was about 0.9 m (3 ft) tall, wore a tight-fitting white suit and had dark skin. Alves was by now convinced that he was seeing devils – he had never heard of UFOs or aliens, and stood frozen with fear. The small men scampered about picking leaves from trees and pulling up bunches of grass.

After a few seconds they clambered back into their craft. The object then rose vertically into the air and flew off in total silence. As soon as it was gone from sight, Alves bolted back home to pour out his story to his local priest and, later, to the local press when reporters came calling.

Whether they were interested in plants, water or were merely passing through, the ufonauts seen in all these encounters largely ignored the humans who saw them. The humans who saw these UFOs and their occupants were often deeply upset by what they had seen, and occasionally suffered physical illness soon afterwards. They could, however, count themselves lucky compared to those whose encounters resulted in the aliens themselves taking a hand.

Rivers and lakes have a particular attraction for some UFOs and their occupants, but quite why this should be so is unclear.

Encounter Casebook No. 5

TYPE>> **Close encounter of the third kind** DATE>> **24 April 1964** PLACE>> **Socorro, New Mexico, USA**
WITNESSES>> **Lonnie Zamora, Sam Chavez**

The Socorro incident was the first encounter that brought together a credible witness, solid evidence and the sighting of humanoids linked to a UFO. It quickly became, and remains, a classic UFO close encounter of the third kind.

On 24 April 1964 at around 5.45 pm Patrolman Lonnie Zamora was heading south from Socorro in his Pontiac police car in pursuit of a speeding motorist across the semi-desert landscape of the area. Another car was heading in the opposite direction but had not yet passed Zamora when an object flew overhead.

The motorist heading north took it to be an aircraft in trouble as it was moving rather erratically. Seeing a police car apparently investigating, the motorist did not report the sighting as a UFO, but merely mentioned the incident when he stopped to refuel his car in Socorro.

Zamora, meanwhile, had not seen the UFO as it passed overhead. He did, however, notice a sudden flash of blue-orange flame in the sky to the west. The bright light was followed by the sound of a roaring explosion. Knowing that a dynamite shack lay in that direction, Zamora feared either that it had exploded or was about to do so. He gave up his chase of the speeding motorist and decided to investigate the apparent explosion.

Zamora turned off the road on to a rough gravel track used only by off-road mining vehicles. As he dipped down into a shallow gully the flame came again. This time Zamora got a better view. It was shaped like a cone with the top narrower than the bottom and was some four times taller than it was wide. There was a plume of dust kicked up from around the base of the flame. The roar came again and lasted for 10 seconds, sounding a bit like a jet but descending from a high to a low pitch.

> The figures were wearing white overalls of some kind and may have had rounded caps or helmets on.

The police car had trouble getting out of the gully on to the top of a low hill. Zamora had to try three times before his Pontiac managed to get a grip on the gravel and ascend the steep slope. As he reached the top Zamora saw an object standing on the ground some distance from the dynamite shack, which was intact.

Zamora at first took the object to be an overturned car and the two figures standing beside it to be a pair of youths who had crashed. The figures were wearing white overalls of some kind and may have had rounded caps or helmets on. They had been apparently talking to each other, but one turned round to look at Zamora as the police car crested the hill.

Zamora then radioed back to his police station 'Socorro Two to Socorro. Possible 10-40 (motor accident). I'll be 10-6 (busy).' The station logged the call, then radioed Patrolman Sam Chavez and sent him out to assist Zamora.

Meanwhile, Zamora had given up trying to drive any further on the rough track. He got out and walked towards the object. The two figures had vanished. He could now see that the object was not an upturned car at all, though it was about the same size as one. It was a whitish-silver colour and shaped like an oval standing on one end. There were four legs projecting down on which the object rested. It was about 1.2 m (4 ft) across and perhaps 6 m (20 ft) tall. On one side was a red marking that resembled an upright arrow within an arch with a horizontal line underneath. There then came two loud thumps as if somebody had slammed a door. Zamora later assumed that this was the sound of the two humanoids shutting their door after they got into the craft.

By this time Zamora was within about 23 m (75 ft) of the object. The roar then began again, getting gradually louder, and the blue-orange flame erupted from the base of the craft. Zamora turned and ran, fearing that the object might explode or the flames engulf him. He struck his leg on his car's bumper and fell headlong, then got back to his feet and ran further trying to keep the patrol car between himself and the object.

Patrolman Lonnie Zamora set out on patrol expecting nothing more exciting than a speeding motorist or two. In fact he was to meet the unknown at close quarters.

Having covered about 7.5 m (25 ft), Zamora glanced back. The object was now hovering on its flame some 4.5 m (15 ft) above the ground. Zamora leapt over the edge of the hill, then turned to look back, thinking that he could duck down into cover at a moment's notice. The flames and the roaring sound had by now stopped and the strange object was hanging eerily motionless. The object then began to move silently off towards the southwest, gathering speed as it went. It stayed only 45 m

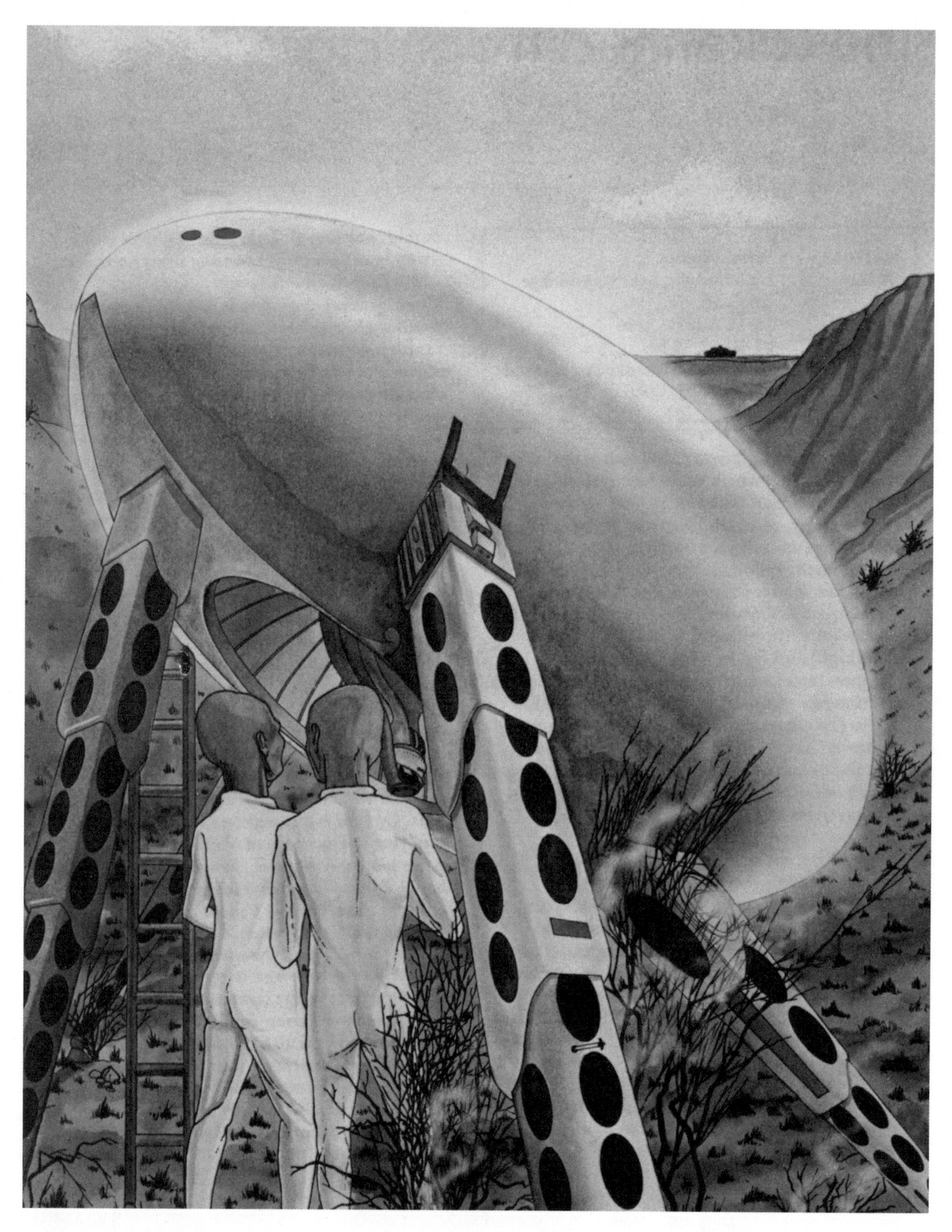

The craft and beings seen by Zamora in an artwork based on his description. This 1964 sighting was the first clear sighting of ufonauts by an impeccable witness.

(150 ft) or so off the ground and vanished behind some hills.

As Zamora was climbing back to his car, Chavez arrived in his patrol vehicle. Together the two policemen went to check the dynamite shack, then turned to investigate the spot where the object had been seen. The greasewood scrub that dotted the area was burnt and smouldering, smoke still curling up from charred twigs. Four prominent marks were seen in the ground where heavy objects had pushed down into the dry soil. Each mark was about 5 cm (2 in) deep and rectangular in shape. Given the soil composition in the area, the object that made the marks must have weighed between 3 and 5 tons. Nearby were smaller, indistinct marks that might have been footprints.

Later analysis showed that the four marks were arranged as if on the circumference of a circle, as they would have been if there were four legs supporting a round object as Zamora had reported. Moreover the centre of the burned area of ground was underneath where the centre of gravity of an object supported on the four legs would have been.

The key importance of the Socorro sighting was that Zamora was an excellent witness who was highly respected by his colleagues. Dr Hynek of Project Blue Book was on the scene just two days later and returned several times to check up on measurements, cross-examine Zamora and investigate further. His conclusion that a real, physical event of an unexplained nature had taken place lifted the reports of ufonauts out of the margins and into the mainstream.

Given the soil composition in the area, the object that made the marks must have weighed between 3 and 5 tons.

CHAPTER 5

Alien Encounters

Although occupants were being seen in, beside and close to UFOs all over the world, there was still little clue as to what the ufonauts were actually doing. They came and went, but nobody seemed to have much idea about why. The best that researchers could say was that the ufonauts must have some purpose to their visits and, perhaps, might one day tell us about it.

Then the ufonauts began to interact with humans and their behaviour began to offer some clues about their intentions. Unfortunately those clues tended to be inconsistent and often contradictory. But then it was rapidly becoming clear that there was more than one type of ufonaut.

The Italians were early starters when it came to interacting with humanoids linked to UFOs. At about 10 pm on 24 April 1950, Bruno Facchini saw what he thought was lightning through the windows of his farmhouse, though no thunder was to be heard. Going out to investigate, Facchini saw a disc-shaped object resting on the ground near a telegraph pole. It was emitting a low but persistent buzzing sound.

Standing around the object were four figures, each about 1.5 m (5 ft) tall. The figures were dressed in tight grey outfits rather like overalls and each wore a transparent face mask which had a tube running from it. One of the figures was holding a tube-shaped device from which came the lightning-like flashes. Facchini thought the figure was using the device to do some repairs or other work on the object.

Facchini watched for some time. The thought then occurred to him that he ought to offer to help, so he stepped forward and called out. The humanoids seemed to see him for the first time and held a hurried conversation in a guttural language. One of the figures then pointed a tube at Facchini. A flash of light blasted out, knocking Facchini off his feet to land a metre or two away on his back. While Facchini scrambled upright and fled, the figures vanished into the object which took off quickly and accelerated out of sight.

Next day Facchini ventured back to the site of his encounter. The ground was marked by scorching and there were indentations apparently made by landing legs.

In 1952, as revealed when the top secret files of the time were later released, the CIA was faced by a bizarre incident that they chose to hush up. A German official named Linke from the Soviet zone of occupied Germany fled to the American zone along with his wife and six children. The reason he gave for fleeing in such haste was that he was worried that the Soviet secret police would come to arrest him after he had stumbled across a top secret Soviet weapon. Instead he had decided to get out fast and reveal what he knew to the Americans in the hope of being allowed to stay in western Germany and, perhaps, be rewarded.

According to Linke, he and his 11-year-old daughter Gabriella had been riding home on his motorbike through a forest near Hasselbacht when one of the tyres burst. As they walked along the road pushing the bike, Gabriella saw something in a clearing. She pointed it out to her father, who at first took it to be a group of deer in the long grass, but then decided it probably wasn't. He left the road and walked off to investigate.

When he reached about 30 m (100 ft) from the object he realized that it was two men dressed in silver-coloured overalls who were bent over and fiddling with something on the ground. One of the men had a light attached to a belt around his waist. Linke walked on until he was just 7.5 m (25 ft) from the men. He then saw, in a position where it was hidden from the road, a large circular object resting on the ground.

Linke then saw, in a position where it was hidden from the road, a large circular object resting on the ground.

Linke estimated that it was about 15 m (50 ft)

German government official Linke mistook a UFO and its crew for a secret Soviet weapon and fled, fearing that the KGB would come to arrest him.

across and shaped like two frying pans placed rim to rim, but without the handles. On top of the object was a cylindrical tower about 3 m (10 ft) tall. Around the rim of the object was a row of dull, yellowish lights about a foot in diameter and spaced some 60 cm (2 ft) apart.

Seeing this object, Linke came to a sudden stop in surprise. His daughter called out to him asking what was wrong. Apparently hearing the girl's voice, the two men looked up and saw Linke. They quickly fled to jump into an opening on the tower and disappeared from view. The lights on the object grew brighter and turned green as a hum began to emanate from it. The tower slid down into the object as the lights turned red and the hum grew louder.

> The humanoid floated up into the air, escaping the man's punches with ease, then it ran off and vanished.

The object began to rise and when it reached a height just above the trees the hum became a whistle that grew in pitch and intensity until it was almost unbearable. The object then flew off at great speed to the north. It had left behind a hole in the ground that was clearly freshly dug.

Concluding that he had stumbled on a military secret, Linke fled to the West. The reactions of the CIA men who had been promised a military secret of great importance only to be given a UFO report can best be imagined. Linke was allowed to stay in the American zone, but his report languished in the files. His encounter with two humanoids dressed in silvery flying suits remained mostly unknown.

On 20 October 1954 a more disturbing event took place near Lake Como. A man had just parked his car outside his house when he spotted a humanoid about 1.2 m (4 ft) tall approaching him. The figure was dressed in a skin-tight one-piece suit of a silken material that glowed slightly. Taken aback, the man stopped and was then hit by a beam of light apparently fired from a belt attached to the humanoid's waist. The man was instantly paralysed. The car keys in his hand began to tingle and moments later the feeling returned to his body. The man was both angry and frightened, so he ran at the humanoid, swinging his fists. The humanoid floated up into the air, escaping the man's punches with ease, then it ran off and vanished.

Although no UFO was seen during this sighting the behaviour of the humanoid was very similar to that of several reported ufonauts, so the incident became part of UFO literature.

In 1961 a new case emerged from the remote jungles of New Guinea that seemed to cast fresh light on the question of UFOs and their crew. The story related to events that had taken place in 1959 at the remote Christian mission of Boaini. The tale took time to get out to the wider world partly because of the remote location and partly because the only literate witness was not entirely convinced that he had seen a UFO at all.

The encounter at Boaini began just after dawn

A helicopter flies over dense forest. New Guinea missionary William Gill at first thought he was seeing a military helicopter, only gradually realizing that a UFO and its crew were visiting his mission.

when Steven Moi, a local convert to Christianity who worked at the mission, saw a disc-shaped object fly overhead. He reported the sighting to the priest in charge, Father William Gill, who dismissed it as probably being an aircraft of some kind.

Five days later, as dusk closed in on 26 June, a bright light was seen by mission worker Eric Kodaware to the northwest. The light came closer, increasing in size until it was about five times the size of a full moon in the sky. Gill was called out and had the object pointed out to him. He in turn called Moi to see if this was the same thing that he had seen. Moi confirmed that it looked similar. By this time 38 people had gathered to watch.

The object came down until it was about 210 m (700 ft) above the ground over the mission's sports field. It could now be seen quite clearly as a round disc with gently curving upper and lower surfaces and a flange around the rim. About 10 minutes after the object first appeared something could be seen moving on its upper surface. When this thing came close to the edge of the rim it could be seen to be a man. He was soon joined by three others. They seemed to be working on a large box of some kind that emitted a blue light. The men came and went several times over the next half hour, seeming to be working at various tasks.

After the men had disappeared for the final time, the object began to rise, climbing through the low clouds. As it passed through the clouds it lit them up, and continued to do so for some time afterwards as if it were hovering just above them. A while later the clouds cleared and various small lights were seen moving about the sky.

At this point Gill believed that he had seen some sort of top secret military aircraft – perhaps a helicopter or hovercraft. Both the Australian and American military had bases in the area and their aircraft regularly flew over New Guinea.

After the men had disappeared for the final time, the object began to rise, climbing through the low clouds.

The following evening the UFO returned at about 6 pm. This time it came down to a height of only about 90 m (300 ft). Again the mission workers gathered to watch. When the men again appeared on the craft, Steven Moi asked Gill if he thought that the aircraft was going to land. 'Why not?' replied Gill, and suggested that they should wave a greeting.

Gill then lifted his arm and waved vigorously as if signalling 'hello'. To his surprise one of the men on the object turned to face him and waved back. Then Moi waved both his arms over his head, and the figure waved back with both arms. The men on the object then disappeared while the craft continued to hover.

It was now 6.30 pm and the mission cook called to announce that dinner was ready. Still thinking that he was dealing with US servicemen, Gill went to dinner, reasoning that he had done all he could to be welcoming, but if the Americans did not want to land it was none of his business.

After dinner, Gill and his team went back outside. The object had moved to hover about a mile away. A short while later it flew off. The following evening some moving lights were seen, but the large UFO did not return to hover over the mission.

It was only when thinking about the events afterwards that Gill thought them odd. He wrote an account of what had happened and sent it to a fellow missionary suggesting that the craft may have been a UFO and asking, 'Do you think Port Moresby (the base of the government authorities) should know about this?'

It took time for the account to reach the authorities, time for an investigation to be launched and longer still for news of the event to filter out to the wider world. In any event, the sighting of humanoids with a UFO by Lonnie Zamora at Socorro (See Encounter Casebook No. 5 on page 86) became well known at about the same time as the Gill UFO. Since Socorro was much easier to reach than Boaini, UFO investigators concentrated on the Socorro case. The events at Boaini only slowly came to become regarded as important as it was the first time that a witness as impeccable as a priest had seen the crew of a UFO, and had interacted with them. The fact that Gill was supported by over 30 locals who had also seen the UFO and its crew only added to his credibility.

... it was the first time that a witness as impeccable as a priest had seen the crew of a UFO ...

In 1964, in the wake of the Socorro incident, an aeronautical engineer from California came forward with his own report and this time there could be no doubt that both the UFOs and their crew had been reacting to him – and in a most unpleasant way. On 5 September Donald Schrum was out hunting in the wilderness area around Cisco Grove with two friends. The three men were archery enthusiasts, and so were armed with bows and arrows.

Schrum became separated from his colleagues and, with dusk drawing on, resigned himself to an uncomfortable night in the open in Granite Creek Valley. Knowing that there were bears in the area, Schrum climbed into a tree and used his belt to strap himself to a wide branch. That way he could doze without fear of falling to the ground and attracting the hungry attentions of a bear.

Police guards on duty at Port Moresby. The authorities there launched an inquiry into the Father Gill sighting, but failed to reach a firm conclusion.

As dusk drew in, Schrum saw three lights flying through the sky overhead, accompanied by a throbbing noise. Taking the lights to be those of helicopters out looking for him, he clambered down from his tree and lit a fire on an area of open land. The 'helicopters' obviously saw the fire as they ceased their random wandering and closed in, starting to circle over Schrum and his fire. As the lights came closer they lost height and Schrum could now see that they were not helicopters, but spinning silvery discs. One of the discs came down very low and seemed to drop two objects. A few seconds later Schrum heard noises coming from that direction that sounded very like something pushing its way through the undergrowth.

Now seriously alarmed, Schrum retreated back up his tree. Two shortish but human-like figures then came into view out of the undergrowth. The figures seemed to be dressed in one-piece silvery costumes that fitted tightly over their slender bodies. Their heads were bald and had odd, bulging eyes rather larger than usual. They inspected the fire, but seemed to shy away from it. They then spotted Schrum and walked over to stand underneath the tree. The humanoids began to shake the tree as if trying to dislodge him.

He tore off another strip of cloth and dropped that, again causing the aliens to temporarily move away.

At this point a third being appeared. This one was rather larger, moved with awkward jerks and seemed to have a metallic skin. Schrum concluded that this was a robot. He shot an arrow at it, but missed. He tried again, but again the arrow fell harmlessly. The robot ignored the arrows and instead moved to join the humanoids.

Schrum was by now very frightened. He recalled that the aliens had not seemed to like the fire, so he tore a strip of cloth from his shirt and set fire to it. He dropped the flaming rag from the tree, at which point the humanoids fled. When the rag burned itself out, the aliens returned apparently more determined than ever to shake Schrum down from his perch.

He tore off another strip of cloth and dropped that, again causing the aliens to temporarily move away. After this fruitless episode had been repeated several times the humanoids seem to lose interest in him and disappeared off into the undergrowth.

The UFOs, which had meanwhile been circling overhead, now picked up speed and sped off. Schrum dozed until dawn, when he clambered down from the tree and warily set off to return to civilization. A few days later he told the story to a science teacher whom he knew, and who persuaded him to talk to the USAF.

Officials working for the USAF Blue Book investigation into UFOs came to interview Schrum about his encounter and took his arrows away for examination. The USAF put the case down as being due to 'psychological causes' and filed it away. The case became public in the 1970s, adding to the by then rapidly growing body of evidence regarding ufonauts.

Equally baffling, though less threatening, was the behaviour of the aliens encountered by William Blackburn at Brand's Flat in Virginia, USA. Blackburn was chopping timber with a large axe in the grounds of the Augusta Country Archery Club when

Forester William Blackburn was chopping timber in Virginia when approached by three ufonauts, one of which tried to talk to him.

he saw two flying objects circling overhead. The objects seemed to be round and slightly flattened.

As Blackburn watched, one of the objects came down, and landed a few metres away from him. A door opened and out stepped three small humanoids. Each figure was about 0.9 m (3 ft) tall and all were dressed in one-piece suits of a shiny metallic substance. Blackburn noticed that there was one finger very much longer than the others on each hand. Then he was struck by the creatures' eyes, which were intensely bright and penetrating.

One of the beings walked over towards Blackburn and began talking to him with some strange, gibbering sounds that Blackburn found utterly unintelligible. After apparently realizing that he could not be understood, the alien turned away and walked back to his companions. All three spoke rapidly to each other before climbing back into their craft and flying off.

Less willing to make contact were the ufonauts seen by a group of nurses at Cowichan on Vancouver Island, Canada, on 1 January 1970. Nurses Dorren Kendall and Frieda Wilson arrived for work at the Cowichan Hospital at 8 am to begin the morning shift. One of their first jobs was to check all the patients in the ward were comfortable, then open the curtains to let in the pale winter sunshine.

As Kendall threw back the curtains of the third floor ward where she worked she was astonished to see a UFO hovering barely 18 m (60 ft) away and slightly below her. The UFO was shaped like a flattened disc with a transparent dome on top of it. Through the dome Kendall could see clearly a pair of chairs facing what seemed to be an instrument panel.

Standing behind the chairs and looking down at something out of Kendall's sight were two men dressed in dark clothing and helmets. So far as Kendall could see the figures appeared to be quite human and about 1.8 m (6 ft) tall.

Through the dome Kendall could see clearly a pair of chairs facing what seemed to be an instrument panel.

Kendall called out to Wilson to come quickly. At that moment one of the men looked up and saw Kendall. He tapped his colleague on the shoulder and began working the controls on the instrument panel. By the time Wilson had reached the window, the object was moving off but was still only about 30 m (100 ft) away. Together the two nurses watched the UFO fly out of sight (as shown in the picture overleaf). Three other nurses also reported seeing the object, though they did not see the humanoids inside.

An altogether less pleasant outcome came from an encounter with ufonauts in Ohio in 1967. On 28 March, factory worker David Morris was driving to his home in Munroe Falls when he saw an odd object in a field. The object seemed to be hovering a few feet above the ground. It was cone shaped, with a base some 3.7 m (12 ft) in diameter and a height of about 7.5 m (25 ft). It glowed with a faint

In 1970 two Canadian nurses in Vancouver got a shock when they drew back a curtain in their ward to find a UFO hovering just yards away.

orange light and had what seemed to be a brighter white globe sitting on top of it.

Morris was so busy looking at the object that he was not really looking where he was going. There was a sudden thump as if he had hit something. Looking forward, Morris saw a short humanoid figure sliding off his bonnet. He slammed on his brakes, thinking that he had hit a child. When his vehicle stopped, he turned around in his seat.

About 9 m (30 ft) behind him he could see a group of humanoid figures slightly less than 1.2 m (4 ft) tall. They were glowing with the same soft orange light as the strange object. The figures were clustered round something lying on the ground. Morris presumed that he had hit one of these creatures and that the body was the object on the floor. One of the humanoids turned to look at Morris. Suddenly terrified of what might happen next, Morris slammed the car into gear and drove off at high speed.

When he inspected his car on the following day, Morris found dents in his front bumper and scratches on the paintwork that he was certain had not been there before the incident. He reported the crash, but a quick investigation showed that no local people had been hurt in a car accident.

Later that year ufonauts that sound remarkably like those that attacked the Sutton family in Kelly, Kentucky (see Encounter Casebook No. 6 on page 120) took a decided interest in two youths near Ririe in Idaho, USA.

On 2 November Guy Tossie and Will Begay were driving south on Highway 26 at about 9 pm. There was an unexpected flash of blindingly bright light from overhead and a flying object appeared ahead of them. The object was disc-shaped and about

A rural road in the USA. These sorts of remote highways are the places where humans – like Guy Tossie and Will Begay – are most likely to encounter a UFO and its occupants.

2.7 m (9ft) wide with a transparent dome on top which measured around 1.5 m (5 ft). The craft glowed with an eerie green colour while orange lights flashed on and off around its rim. Within the dome could be seen two figures. Begay's car came to a gentle halt, its engine and lights cutting out as it approached the UFO.

The head had huge, pointed ears flapping on either side and bulbous eyes fixed to the side of the head.

The dome of the UFO flipped open and one of the figures jumped out, seeming to float down to the ground. It approached the car and opened the driver's side door, causing Begay to squirm across to join Tossie on the passenger's side. The ufonaut got into the car and sat at the driving wheel without touching it. The UFO then began to move off, apparently towing the car somehow as it began to trundle off the road and into a recently harvested wheat field.

The two young men were staring in fear at the creature in the car with them. They described it later as being about 0.9 m (3 ft) tall with spindly arms and legs but a very large, domed head. The head had huge, pointed ears flapping on either side and bulbous eyes fixed to the side of the head. The mouth was a thin slit and the nose was tiny. It appeared to be naked.

As the car bumped over the stubble, Tossie opened the door and fled. He ran as fast as his feet could carry him, glancing over his shoulder to see that one of the creatures, carrying a light, was chasing him. Spotting a house, he leaped the garden fence and hammered on the door. The door was opened by farmer Willard Hammon who would later describe Tossie as being so terrified that he could not talk.

After some minutes, Tossie calmed down enough to tell Hammon what had happened. Uncertain what to believe, Hammon set off to look for the missing Begay. Hammon discovered the car standing with its engine running and lights on in the middle of a field. Begay was cowering on the back seat with his eyes firmly shut and arms crossed over his face. Neither the aliens nor the UFO were anywhere to be seen.

Hammon managed to persuade Begay that the little creatures had gone, and led him back to the farmhouse. There Begay said that after Tossie had fled, the intruder had turned to face him and had begun talking to him in an aggressive way as if it were angry. Begay had been unable to understand a word of what was said, describing the speech as being like that of a twittering bird. It was at this point that he had shut his eyes, expecting to be attacked at any moment. Instead he had heard the alien leave the car and walk off.

Entirely uncertain what to believe, but being able to see that the two young men were terrified by what had happened, Hammon called the police. The police investigation found that a neighbouring farmer had seen a light moving low through the sky and that his cattle had bolted, some minutes before the encounter took place.

Unwilling to make contact was the humanoid seen out collecting desert plants near Nazca, Peru, on 3 February 1972.

Tito Rojas and Adolfo Penafiel ran an animal feed business and were driving their truck north over the semi-desert landscape to a ranch near Nazca. They had reached a flat plain known as the Pampa Carbonera when the truck's radio was swamped with static. A few seconds later the engine began to misfire, then cut out completely.

As the truck coasted to a halt, the two men got out. Rojas went to inspect the engine, while Penafiel glanced idly over the desert landscape. He saw what he thought was a car parked some distance away amid the desert scrub. Thinking that it was another car in trouble, Rojas and Penafiel walked towards it to offer assistance.

As they drew closer the two men realized that the object was no car. It was an oval metal craft about 15 m (50 ft) long and 4 m (13 ft) tall that was resting on three thick legs. The craft's surface was of a brightly-polished silvery metal without any obvious joints, welds or even openings.

As the two men stared at the craft, a figure came walking into view (see illustration overleaf). Dressed in green overalls and with a helmet covering its head, the figure was moving slowly and staring at the desert plants. It stopped now and then to bend closer or to pluck a few leaves and twigs. As soon as it saw Rojas and Penafiel,

> Whatever the reason why the alien had been collecting plant samples in Peru, he did not complete his mission.

the humanoid stopped in apparent surprise. Then it turned and ran away, quickly disappearring around the rear of the object.

Seconds later the object began to emit a whine that grew rapidly in pitch and power. It then rose vertically and disappeared into the cloudless desert sky. After discussing the bizarre event for some time the two men returned to their truck, which started instantly. They then hurried into Nazca to report the encounter.

Whatever the reason why the alien had been collecting plant samples in Peru, he was interrupted and did not complete his mission. Also interrupted in their task – whatever it might have been – were the three ufonauts who were seen on the Isla de Lobos off the coast of Paraguay on the evening of 28 October 1972.

The island is home to a small base of the Paraguayan Navy which maintains a lighthouse and various military radars and navigational equipment. There is a garrison of five men which stays on the island for up to a fortnight before being relieved.

On 28 October Corporal Juan Figueroa had the task of periodically checking the generators that powered the lighthouse. At 10.15 pm he left his

Tito Rojas and Adolfo Penafiel were driving in a remote desert area of Peru when they came across this UFO and its helmeted occupant.

colleagues playing cards in the guardhouse to walk about 45 m (150 ft) to the hut where the generator was housed.

As he left the guardhouse, Figueroa noticed some dull, flashing lights beside the generator hut. He knew of nothing that should emit such lights and hurried forward in case something was wrong. He was about halfway to the hut when he was able to make out that the lights were attached to a large, dark object resting on the ground beside the hut. The object was shaped like an upturned bowl about 5.5 m (18 ft) across, 3 m (10 ft) tall and with a dome or bulge on top of it.

When about 18 m (60 ft) from the object, Figueroa realized that there was one man standing next to it and a second climbing down what seemed to be a ladder to join the first. Judging their height by that of the shed, Figueroa estimated that the figures were a little less than 1.5 m (5 ft) tall. A third figure, rather taller than the other two, now emerged from the dome and began to climb down. All three figures were dressed in dark, tight-fitting

> **He realized with alarm that he was totally paralysed.**

clothing and had their heads covered with helmets or hoods of some kind.

Realizing that he had intruders on the military base to deal with, Figueroa drew his pistol and shouted out a challenge. The three figures turned to face Figueroa and the corporal felt a vibration or mild electric shock run through his body. He realized with alarm that he was totally paralysed. A powerful voice inside his head seemed to be telling him not to shoot.

The three figures quickly climbed back inside the craft and shut a door behind them. The craft then rose into the air with a humming noise and hovered briefly at a height of about 36 m (120 ft). It then tipped upwards, emitted a fireball and shot off at astonishing speed.

Freed from his paralysis, Figueroa raced back to the guardhouse. His comrades were alarmed as he ran in shouting excitedly and waving his pistol about barely a minute after he had left. Sub-Officer Cascudo sent two men out to mount a quick patrol while he tried to get some sense out of the corporal.

The guards found nothing, and when Figueroa managed to tell his story they were inclined to think he had been dreaming. However, Figueroa insisted on making a formal report and was later summoned to tell his story to senior officers. At first Figueroa was inclined to think that he may have foiled a plot by an enemy country, but no explanation was ever given to him so he eventually went public.

Another incident at a military base took place at Alsworth Air Force Base in South Dakota, USA, on 16 November 1977. At about 9 pm, an automatic alarm on the northern section of the perimeter fence was triggered. Two soldiers, named Jenkins and Raeke, were ordered out in a patrol vehicle to investigate. Expecting to find nothing more exciting than the usual sensor malfunction or some local wildlife, Jenkins and Raeke were surprised to see a bright light just outside the perimeter fence close to where the alarm had been activated.

A generator shack much like the one beside which Corporal Juan Figueroa encountered a UFO and its crew at a Paraguayan naval base in 1972.

Roswell 1947. The initial press interest in what may have been a UFO crash was stifled by this photo showing clearly that the 'wreckage' was from a weather balloon. However, it is now widely accepted that these pieces of material were not those found at the crash site.

Jenkins stayed with the vehicle while Raeke went off on foot to check the fence. As he reached the edge of a slight slope, Raeke saw a figure walking towards the fence from the centre of the base. The figure was about the size of a man, dressed in green overalls that had a metallic sheen to them and had what looked like a motorcycle helmet with visor over his head.

Raeke shouted a challenge, which was ignored. He then raised his rifle and shouted a second challenge. The figure turned towards Raeke and aimed some sort of a box in his direction. There was a sudden blast of light. Raeke was thrown backwards to land some feet away. He could not find his rifle and his hands were covered in severe burns.

Hiding behind some bushes, Raeke got his radio working and alerted Jenkins to the fact that he was under attack by an intruder. Jenkins passed the alarm on to the main security control and set off to find his comrade. He found Raeke and led the shaken soldier back to the vehicle. Jenkins then set off to look for the intruder himself.

Having already seen his comrade wounded, Jenkins was in no mood to take chances. When he saw two figures in the green costume described by Raeke, Jenkins took cover before shouting an order for them to halt. The figures did not stop, so Jenkins opened fire. The figure at which Jenkins aimed was thrown forward as if he had been shot in the back, the other dropped instantly to the ground. Seconds later a bolt of light streaked out, but missed Jenkins. From cover Jenkins watched the two figures creep away.

> ... other documents relating to the incident have remained classified as secret.

Moments later a green disc, estimated by Jenkins to be about 6 m (20 ft) in diameter, rose into the sky from the direction in which the intruders had disappeared. It hovered briefly, then flew off at high speed.

This particular incident is shrouded in some mystery. The security report involving the actions of Jenkins and Raeke was released by the US military some years later. However, other documents relating to the incident have remained classified as secret. A medical report showed that Raeke made a surprisingly swift and complete recovery from his burns.

The reluctance of government officials to speak openly about the various UFO sightings and apparent alien encounters experienced by military personnel has long been a great frustration to UFO investigators. In no case has this caused more friction, distrust and confusion than in what may well be the most famous alien encounter of them all: Roswell.

The Roswell incident took place in early July 1947, but at the time attracted little attention. On 8 July, a press release issued from Roswell Air Force Base by Lieutenant Walter Haut stated that a flying saucer, as UFOs were then known, had crashed near Roswell and the wreckage had been recovered by the military. A few hours later a second press release was issued, stating that the wreckage was actually from a weather balloon. The mistake was explained as an error on the part of the personnel from Roswell who were unfamiliar with the new type of weather balloon that they had found. The air force displayed pieces from a weather balloon. The press promptly lost interest in the story.

The story was to remain lost for decades. It was not until 1977 that anyone took any new interest in the events at Roswell, and then it was entirely by accident. Stanton Friedman, a nuclear physicist who took a part-time interest in UFOs was appearing on a local TV show in Louisiana. Before going on to talk about his UFO investigations, Friedman was chatting to the station manager. The manager gave Friedman the address of a local man named Jesse Marcel who was known in the district for his claims to have once had very close contact with a UFO.

For years, Marcel had remained unhappy with the weather balloon story but had been unable to prove anything.

Friedman duly contacted Marcel, who turned out to be the one-time Major Jesse Marcel who in the late 1940s had been intelligence officer at Roswell Air Base. Marcel told Friedman that one Independence Day weekend he had been sent out from Roswell to a nearby site to collect some debris that he was told had come from a flying saucer. He had done so and was ordered to take the wreckage to Wright Field, now the Wright-Patterson Air Force Base.

Marcel said that a press release had been issued, but that it was later quashed. He was stopped en route to Wright Field and ordered to hand over the wreckage to a senior officer, which he did. He was later told that the wreckage had actually been from a new type of weather balloon, but was unconvinced. For years, Marcel had remained unhappy with the weather balloon story but had been unable to prove anything.

Friedman was interested in the story, but unfortunately Marcel could not recall which year the event was supposed to have happened nor did he have contact details for any of the other service personnel he said had been involved.

Friedman contacted his fellow UFO researcher William Moore for help. Moore happened to have archive copies of a magazine called *Flying Saucer Review.* He began trawling through old copies and in a 1955 edition came across an interview with the British TV personality Hughie Green in which he mentioned that during a 1947 trip to the USA he had seen newspaper reports about a downed UFO being captured by the military, but had heard no more about it and so assumed that it had been a mistake.

By putting the year mentioned by Green together with the date mentioned by Marcel, Friedman and

A contemporary illustration shows rancher Mac Brazel inspecting the strange wreckage on his land. In fact the metal fragments did not carry American lettering as shown here, but other symbols that nobody could identify.

Moore now knew that they were looking for reports of a UFO crash near Roswell in early July 1947. They got hold of archive copies of the local newspapers and soon had the names of the rancher on whose land the wreckage had been found, the local sheriff who initially handled the investigation and several military personnel involved. They set off to track those people down and see what they had to say.

Such were the beginnings of the Roswell investigation, which since then has taken on something of a life of its own. By 1980 Friedman and Moore had interviewed 62 people. They were by then confident that something very strange had happened at Roswell and that the USAF had gone to great lengths to cover it up. The most likely explanation, they thought, was that a UFO had crashed and its wreckage recovered.

They began publishing their findings and in 1986 the NBC television show *Unsolved Mysteries* broadcast a programme based on their research. It was this programme which brought the Roswell incident to the attention of the general public and to the wider community of UFO researchers.

> It did seem ... that the clearout of Roswell's 1947 files had been rather more thorough than mere routine would demand.

In 1989 the story took a dramatic new twist when Glenn Dennis, who had worked as a civilian mortician in Roswell at the time, came forward to say that he had been told that bodies had been recovered from the crash. The air force medical team, he claimed, had phoned him to ask him detailed questions about how to preserve bodies and that he had later been threatened by the military to keep quiet.

When news of Dennis' story became public an air force photographer, who preferred to remain anonymous, came forward to announce that he had seen and photographed the bodies.

In 1994 the congressman for New Mexico, Steven Schiff, took up the story. He asked the General Auditing Office (GAO) to investigate the official government files relating to the incident at Roswell in 1947. This led to two reports.

The first report was from the GAO, stating that most of the records from Roswell Air Force Base relating to 1947 had been destroyed. Such destruction of files is not unusual. The government has only limited space to store documentation and routine paperwork detailing food bought and served, sick lists, pay lists and the like is often simply thrown away after a number of years. It did seem, however, that the clearout of Roswell's 1947 files had been rather more thorough than mere routine would demand.

The second report came from the USAF. This confirmed that wreckage had been found at Roswell in early July 1947 and admitted that it had not been from a weather balloon. The second press release, it stated, had been a cover-up. The real story, this USAF report said, was that a top secret Mogul craft had crashed. These Mogul craft were ultra-high altitude balloons (see photograph on page 113), carrying highly advanced electronic equipment able to detect and analyse nuclear explosions wherever on Earth they took place. In effect, they were spy aircraft sent up to discover the secrets behind the Soviet atomic tests then taking place. The USAF report explained that in 1947 the need to maintain the secrecy of the Mogul flights had been paramount, and so a cover-up had taken place.

Since then the Roswell story has moved on, but the essentials have not. Numerous witnesses have testified to the strange events of those few days in 1947. If all of them are to be believed about everything they say, then it is beyond doubt that a UFO crashed at Roswell, that the craft was recovered mostly intact by the USAF and that four dead aliens were found in the wreckage. However, the USAF itself stands by its story that the only object to crash near Roswell in July was a Mogul.

It is very difficult to sort out the truth about what really did happen at Roswell. Most of the witnesses were speaking forty or more years after the event. Human memory is a notoriously unreliable thing. Incidents can be forgotten or exaggerated over time. Details can be lost or blurred. Events that took place some days or even weeks apart can become conflated and so are later recalled as if they happened at the same time. Moreover some of the evidence is not first hand. It relates not to what a witness saw himself but to what somebody else – who has since died – said that they saw.

That said, many of the witnesses were aware at the time that what they were seeing was very out of the ordinary and so they are more likely to recall details accurately.

In the final analysis there are some facts about Roswell that can be stated with certainty and others that remain disputed and controversial.

Years after the initial reports from Roswell in 1947, word began to leak out that a UFO had crashed and alien bodies had been found within. A top secret military operation had apparently cleared away the wreckage.

What can be said with some certainty is that something unusual happened in and around Roswell. The sequence of events has now been fairly clearly established as follows.

On Tuesday 1 July the radar stations at Roswell Air Base and nearby White Sands, a missile testing base, picked up a UFO. It flew over very quickly and performed manoeuvres impossible for any known aircraft. It was not seen visually. The following day civilian residents of Roswell, Mr and Mrs Dan Wilmot, were sitting on their porch at around 10 pm when they saw an object flying west at high speed. They later described it as being shaped like two saucers joined around the rim.

On the evening of Friday 4 July a radar-visual sighting of a UFO took place over Roswell while a thunderstorm raged. Local resident William Woody saw a white UFO with a red trail behind it flying northwards over Roswell. Rancher W. 'Mac' Brazel heard a loud explosion over his ranch, but could not see anything. Further north, Jim Ragsdale and Trudy Truelove were camping when they saw a craft fly overhead, apparently in trouble, and appear to crash a mile or so away near Corona. Unfortunately it has proved impossible to discover if these three reports relate to the same object. The witnesses could

Hanger 84 at Roswell Air Base. It was to this hangar that the wreck of the crashed UFO was allegedly taken amid great secrecy in 1947.

recall only that they happened late at night.

Meanwhile the Roswell radar had again picked up an unidentified craft flying nearby. That object vanished abruptly, indicating either that it had crashed or that the trace had been a malfunction or one of several possible natural blips.

Early the next morning Brazel rode out in the direction from which he had heard the explosion. He found strange debris scattered over an area about 630 m (2100 ft) long by 180 m (600 ft) wide. Most of the debris consisted of a thin, silver-coloured film that was both extremely light and very strong. There were also pieces of lightweight struts with an I-shaped section. Again these were very strong, but some of them had odd markings which looked a bit like some type of hieroglyphic writing. Brazel collected up some of the debris and showed it to his neighbours.

Meanwhile, Glenn Dennis, the mortuary assistant, took a series of calls from Roswell Air Base asking for advice on preserving bodies. From what was said, Dennis got the impression that somebody important had been killed in an aircraft crash and that attempts were being made to make the mangled body look presentable. Then the military sealed off an area of land north of Roswell, near Corona.

On Sunday 6 July Brazel took the pieces of debris to Sheriff George Wilcox, who called the Roswell Air Base to report the find. He seemed to think that if anything odd had fallen from the sky it was most likely something to do with the air force. Colonel William Blanchard, the base commander, sent Jesse Marcel to Wilcox's office to interview Brazel and collect the pieces of debris. The following day Blanchard sent a team of men to Brazel's ranch to pick up all the pieces of debris and take them to Roswell Air Force Base.

...the mortuary assistant took a series of calls from Roswell Air Base asking for advice on preserving bodies.

On Tuesday 8 July, Blanchard told Lieutenant Walter Haut to write up a press release about the crash on the Brazel ranch. He was told to ascribe the wreckage to a 'flying disc', which is what he did. The press release went out at 2 pm local time.

Meanwhile, Marcel had set off for Wright Field with the wreckage bundled up in packing cases on board a B29. The aircraft made a stop at Fort Worth at about 4 pm and Marcel was told to report to Brigadier General Roger Ramey. Ramey announced that the wreckage was no longer Marcel's responsibility and that he could go back to Roswell. Marcel handed over the packing cases and left.

At 5.30 pm Major E. Kirton issued a new press release stating that the 'flying disc' of the earlier press release was a mistake. The debris had in fact come from a weather balloon. The FBI and air force then both moved quickly to repeat this version of events and so kill the story so far as the media was concerned.

The next day, 9 July, more military personnel arrived at Brazel's ranch to swarm over the field of debris, picking up even the smallest fragments. Brazel himself was interviewed repeatedly by the military who demanded to know every last detail of what he had seen and found. He was eventually released, though not before being instructed in very stern terms not to tell anyone anything.

This much is agreed by almost everybody who has ever been involved in the Roswell incident. Most of it can be gleaned from contemporary accounts that have never been challenged. Most of it is even acknowledged by the USAF.

According to the USAF version of events, the bang heard by Brazel and the objects seen by others was the Mogul balloon exploding and showering its remnants down on the Brazel ranch to form a field of debris. The wreckage Brazel found was the remains of the Mogul. Neither Marcel nor Blanchard had ever seen a Mogul nor even knew of its existence, so they were at a loss to identify the wreckage.

According to the USAF version of events, it was not until 8 July that the air force had linked the reports coming in from Roswell to the fact that a Mogul had malfunctioned and gone missing. It was when the link was made that Ramey stepped in to take the wreckage from Marcel and Kirton spread the false story about a weather balloon. The subsequent efforts to clear the Brazel land of debris, to clamp down on news and to silence Brazel himself are explained as efforts to keep the top secret Mogul project secret.

In itself the USAF version makes sense, but it ignores the sighting of a flying object in trouble by Ragsdale and does not explain why a stretch of land some miles north of the Brazel ranch was sealed off for several days.

Brazel himself was interviewed repeatedly by the military who demanded to know every last detail of what he had seen and found.

The answer to those riddles may lie in the testimony that has been gathered in the years since Marcel first spoke to Friedman and Moore. Some of this evidence comes from apparently reliable witnesses, some from less certain quarters. Some of it is first hand; other elements are second or even third hand. Despite these limitations, most of this later evidence does fit together and when combined and correlated produces a fairly coherent timeline of events.

Under this scenario, the explosion heard by Brazel was a UFO suffering a severe malfunction, perhaps being struck by lightning. This explosion caused parts of the UFO to shower down into the debris field. The UFO then flew on northwards erratically and probably out of control to crash into a hillside off Pine Lodge Road near Corona. The radar trace gave the authorities at Roswell the approximate site of the crash and next morning personnel were sent

A Mogul spy balloon. In 1947 these high-altitude devices were used to spy on Soviet atomic weapons tests and were highly classified. The USAF has explained the Roswell incident away as being nothing more than the crash of a Mogul balloon.

out to secure the area and search it. The crashed saucer was found at around 7 am. The military did not know of the debris field on the Brazel ranch until they got a phone call from Sheriff Wilcox, whereupon Marcel was sent to temporarily seal the area. He would later be replaced by more trusted personnel.

Various witnesses have testified to seeing a crashed object near Corona, others to seeing dead bodies of humanoid creatures about 1.2 m (4 ft) tall. Accounts differ somewhat as to the size and shape of the crashed UFO, and to the appearance of the aliens. The discrepancies are not, however, too serious and could easily be explained by different viewpoints and the years that have passed since.

There is ... evidence that the US Government is not telling all it knows about UFOs and alien contacts.

The UFO is generally thought to have been a dull silver colour and to have been shaped like a crescent rather than a saucer. Some suggest that there was a central fuselage. The aliens are described as being basically humanoid. The heads were large in proportion and were bald with highly domed foreheads. The eyes were large, the noses small and the mouths mere slits. The wreckage and bodies were reportedly taken to Wright Field for future study.

Again, the story is consistent and fits the known facts. But as with the USAF version, there are problems. Leaving aside the unreliability of reports made thirty or forty years after the event, some of the most dramatic elements of the story cannot necessarily be tied directly to Roswell in early July 1947 and may relate either to other events or be open to different interpretations. There is also the problem of what has happened to the UFO and its occupants since. If the USAF had a highly advanced spacecraft from another planet to study it might be expected that after all these years they would have fathomed at least some of its secrets. To date, however, the technology of the USAF has been tediously Earth-based and explicable, though at times spectacular.

There is, however, a growing and impressive amount of evidence that the USAF was covering up something very important at Roswell, and has continued to do so ever since. The Mogul project was declassified a few years after the crash, but the incident at Roswell remained top secret. Why, researchers wonder, was the incident kept quiet if the project it was supposedly protecting was already in the public domain?

There is, of course, other evidence that the US Government is not telling all it knows about UFOs and alien contacts. Persistent rumours and fragmentary evidence continues to leak out indicating that something very odd and unusual is being kept at Wright-Patterson Air Base.

Whatever the truth may be regarding Roswell, by the late 1980s UFO researchers had amassed some impressive evidence indicating that aliens were visiting Earth. They were trying to make sense of the apparently bewildering variety of aliens that had been reported. It must be remembered that the

> ... researchers gradually came to realize that most sightings of aliens fitted into one of a small number of types.

people viewing ufonauts almost all did so for only short periods of time and were, quite naturally, both surprised and shocked by what they were seeing. In the circumstances it is hardly surprising that descriptions of aliens varied.

Police and lawyers who interview witnesses to crimes are well aware of the fact that a witness, even an observant and entirely truthful one, will report only a part of what he or she has seen. People will often notice one particular feature, say that a man was wearing a shirt identical to the witness's own, but will not recall another feature, such as the man's hair colour. Added to that are the facts that witnesses will often have their view partially blocked by a bush or other object, or may be largely concerned with fleeing rather than taking detailed notes.

It is hardly surprising that details vary between different reports. However, by analysing the various sightings and statements, researchers gradually came to realize that most sightings of aliens fitted into one of a small number of types.

Despite strenuous denials from the US military, reports and rumours persist that they have captured at least one living alien, and possess several other dead bodies recovered from one or more crashed UFOs.

Among the first to be reported in any numbers are the aliens generally known as Nordics. These aliens were given this name as they tend to be very human-like and almost invariably rather tall with long blond hair and blue eyes, similar to the popular image of Scandinavians. They are usually seen dressed in one-piece, tight outfits that are described as being like ski suits or motorbike leathers.

In the 1960s several Europeans reported encountering aliens that were very human-like in appearance and who came to Earth with good intentions. This type of alien has been dubbed 'Nordic'.

The Nordics were generally reported to behave in an aloof or uninterested fashion. They would stare at the humans who saw them and sometimes seem to take notes or talk about the humans. Some witnesses reported that the Nordics seemed friendly. Reports of the Nordics were most frequent during the 1950s and 1960s, but declined in number during the 1970s and have never really become numerous again.

> By far the most common type of alien to be reported is the so-called Grey.

Very different were the Goblin types. These aliens were short, usually around 0.9 m to 1.2 m (3 ft to 4 ft) tall. Although they were humanoid in that they walked upright on two legs and had two arms and one head, they were otherwise bizarre in appearance. They generally had long arms that ended in claws or talons, large heads with grotesquely pointed ears and eyes that were reported to look evil or malevolent. Some Goblin types were reported to be covered in dense hair or long fur, but most were naked with a smooth or reptilian skin.

The behaviour of the Goblin types was as aggressive as their eyes would indicate. They were prone to attack humans, inflicting cuts and scratches, and sometimes seemed intent on dragging the human into their spacecraft. They were usually, but not always, said to be much stronger than their size would suggest they should be. There was usually a noticeable lack of any technology or implements when compared to other reported alien types. Perhaps fortunately these Goblin types have always been rare and are not seen often.

Seen rather more often than either the Nordics or the Goblins are the Tricksters. These aliens appear to be broadly similar to humans, though they rarely stand more than 0.9 m (3 ft) tall. They are often said to wear one-piece suits and sometimes have on helmets or face masks. These Tricksters are often reported to be deeply interested in plants and animals: very often crops or domestic livestock. They have been seen to take samples from plants and attempt to catch animals.

The Tricksters, like the Nordics, seem somewhat uninterested in humans and react as if the arrival of a human witness is a bit of a nuisance. Sometimes they run away, or take off in their UFO. At other times they will motion for the human witness to leave. If that does not work, they may paralyse or incapacitate the human, often with a tool that emits a beam of light.

By far the most common type of alien to be reported is the so-called Grey. These beings stand just under 1.2 m (4 ft) tall and are humanoid in appearance. They have thin, almost spindly arms, legs and bodies, but very large and rounded heads. The arms and legs are sometimes said to lack elbows or knees and to end in long fingers and toes that are opposable, but again lack clearly defined joints. The heads are hairless, and often earless and noseless. The mouths are usually described as being mere slits, if they are mentioned at all. It is the eyes that seem to be the most noticeable features of this type of alien. They are invariably described as being extremely large, jet black and almond-shaped. Some witnesses report them to have hypnotic powers, while others believe that any telepathic communication they receive is by way of these mesmeric eyes.

Attacks by Greys are generally more sophisticated than the physical violence of the Goblins, but nonetheless disturbing.

The behaviour of the Greys is generally unfriendly. Like the Tricksters they sometimes take an interest in plants and animals, but unlike the Tricksters they are very interested in humans. Attacks by Greys are generally more sophisticated than the physical violence of the Goblins, but nonetheless disturbing. They will try to induce humans on to their craft for a variety of unpleasant purposes including, as we shall see in Chapter 7, abduction and medical experimentation.

A few witnesses have seen Greys accompanied by taller aliens. These tall Greys are around the height of a human and seem to be in command. Some witnesses describe them as being tan, rather than grey in colour.

The Greys were first reported in North America, being seen in small numbers during the 1950s and 1960s. During the 1970s sightings of the Greys increased dramatically and by the 1990s there were more sightings of Greys than of all other types of aliens put together. They remain the most regularly reported by witnesses.

In addition to these basically humanoid aliens, there are other types that owe little or nothing to a human-like appearance. Robots are, perhaps, the most numerous of these ufonauts. They tend to be of metallic appearance and often have flashing lights attached to their bodies.

A final category goes by various names, such as Exotic or Bizarre, depending on the researcher involved in categorizing them. The category includes all sorts of odd and unusual alien reports. Some witnesses have seen disembodied brains, headless birdmen or bouncing blobs of jelly. Such aliens are rarely seen and a great many are reported only once.

The abiding mysteries linked to all these aliens were:

Where were they from?

What did they want?

UFO researchers were at a total loss to explain what the aliens were really up to. What was needed, it was felt, was for the aliens to get in touch with humanity and tell us.

Very soon they began to do exactly that.

The most often reported type of alien is the 'Grey'. This type is rarely friendly to humans and has sometimes proved to be distinctly hostile.

Encounter Casebook No. 6

TYPE>>Close encounter of the third kind DATE>>21 August 1955 PLACE>>Kelly, Kentucky, USA
WITNESSES>>Elmer 'Lucky' Sutton, Vera Sutton, Glennie Lankford, Bill Taylor, three other adults and four children

When the news of what became known as the Kelly-Hopkinsville encounter got out nobody – not even seasoned UFO investigators – quite knew what to make of it. The appearance and behaviour of the aliens involved was quite unlike that of most reported ufonauts, though there were a few obscure cases that mirrored the experiences of the unfortunate Sutton family. Since that terrifying night in 1955 others have had similar experiences, but they remain few and far between.

The evening began with what might be described as a fairly routine UFO sighting of the daylight disc type, but soon became something very much more dramatic and unusual.

Bill Taylor was visiting his neighbours, the Sutton family, at their remote farm outside the settlement of Kelly near the town of Hopkinsville. At around 7 pm Taylor went out to the farmyard to get a bucket of water from the well. As he returned to the house he saw a strange disc-shaped UFO flying low some distance away. It trailed an exhaust that seemed to be made up of gases of various colours. As Taylor watched the UFO halted, then floated down to disappear behind a line of trees and apparently land in a dried-up river bed.

Taylor hurried back into the house to tell the six adults and several children of the Sutton family about what he had seen. They did not believe him, and Taylor dropped the subject.

About an hour later the farm dogs began barking loudly. Taylor and Elmer Sutton left their supper to go to the kitchen and look out across the farmyard. They both saw on the far side of the yard a most bizarre apparition. The figure they saw was about 0.9 m (3 ft) tall, walking upright on short legs and with very long, ape-like arms. The creature's head was disproportionately large with enormous pointed ears, bulging eyes and a slit-like mouth. Even more unnerving was the soft silvery glow that it emitted from its body. The being was joined by several more that then began to wander apparently aimlessly around the farmyard.

The figure they saw was about 0.9 m (3 ft) tall, walking upright on short legs and with very long, ape-like arms.

Elmer and Taylor, being robust Kentucky farmers, picked up their guns and emerged from the kitchen doorway. Elmer shouted a challenge to the beings. One of the aliens then turned towards the two men

The aliens that attacked the Sutton family farm seem to have emerged from the UFO that neighbour Bill Taylor had seen landing nearby a short time beforehand.

and ran at them with its arms held above its head. Elmer Sutton did not hesitate. He raised his shotgun to his shoulder and blasted the creature at point blank range. The impact of the shot knocked the creature flying backwards. It lay motionless on its back for a moment, then scrambled to its feet and ran off, seemingly uninjured.

The impact of the shot knocked the creature flying backwards. It lay motionless on its back for a moment, then scrambled to its feet and ran off, seemingly uninjured.

The two men went back into the house and slammed the door behind them, while the rest of the Sutton family crowded into the kitchen to see why the shot had been fired. Peering out of the windows, Taylor said he thought the creatures had gone, but then footsteps were heard on the roof of the house.

Once more Taylor and Elmer ventured out with their guns. As Taylor stepped out of the door one of the aliens grabbed his hair in its long, spindly hands. Whipping round, Taylor shot the alien with a .22 rifle. It flipped backwards over the roof, then seemed to recover and fled into the darkness. Elmer had meanwhile spotted one of the creatures squatting in the branches of a tree. He shot it, watching it jerk back from the branch then float gently to the ground before it too ran off.

After terrifying the Suttons and causing much mayhem and confusion, the aliens that launched the assault slipped quietly away into the woods towards their landed UFO.

Once more the men retreated to the house, this time barricading the doors and windows. For the next three hours the family remained – terrified – inside the house while the men took potshots at the aliens whenever they appeared. The family noticed that when hit by bullets or shotgun pellets, the aliens made a sound rather like a stone being thrown hard into an empty metal bucket. The strange glow of the creatures would increase whenever a shot was fired, only to fade quickly back to normal.

In the small hours of the morning the family and Taylor held a hurried conference. With no way of knowing if the aliens were still about or if other farms and villages were also under attack, the Suttons and Taylor decided that staying put was not an option. They decided to head for Hopkinsville Police Station in the hope that the stoutly built station containing armed men was holding out.

The women and children raced to the cars, while the men kept guard with their loaded guns. Once in the cars the family raced to Hopkinsville to find Chief of Police Russell Greenwell enjoying a very quiet and uneventful night in the company of his deputy George Batts.

The police found it hard to credit any truth to the story, but Greenwell knew Elmer Sutton well as a calm, stolid and tough farmer. Greenwell later commented: 'Something frightened these people. Something beyond their comprehension.' Believing that the terrified Suttons and Taylor had encountered something real, Greenwell summoned the four policemen on duty and together they all set off for the Sutton farm.

All was quiet when the convoy arrived. There was no sign of the aliens or of the UFO when the river bed was explored. The farmhouse and farmyard, however, did bear the marks of intense battle. Windows were smashed, doors unhinged and surrounding trees and outhouses bore the clear marks of the shots fired during the siege.

Next day the police alerted the local WHOP radio station and reporter Bud Ledwith drove out to the Sutton farm to investigate. Ledwith found that the men were out working, so interviewed the women, including the family grandmother Glennie Lankford. They told him all about the events and helped him to produce a drawing of the creatures. When the men

There was no sign of the aliens or of the UFO when the river bed was explored. The farmhouse and farmyard, however, did bear the marks of intense battle.

returned, Ledwith interviewed them separately. They produced an identical story and picture.

Over the days that followed the story spread rapidly, and gained in the telling. Soon hundreds of people were flocking to see the site of the 'great alien battle'. Understandably the Sutton family quickly tired of having strangers wandering over their farm damaging crops and taking photos. Elmer blocked access and chased off sightseers with threats.

Despite coming in for endless teasing, some unpleasant comments and a good deal of ridicule, the Suttons and Taylor never changed their story. They remained convinced that they had been under attack that night and had been lucky to escape with their lives.

Encounter Casebook No. 7

TYPE»**Close encounter of the third kind** DATE»**1 July 1965** PLACE»**Valensole, France** WITNESS»**Maurice Masse**

The encounter with aliens experienced by French farmer Maurice Masse did not attract a huge amount of attention at the time, at least not outside the area where he lived. However it has since been recognized as one of the most crucial early sightings of aliens as it established a number of patterns that were to be repeated time and again.

The day of the encounter began normally enough for Masse. As a farmer he was used to working long hours, so he rose at 5 am, ate a quick breakfast and climbed into his tractor to start work in his lavender fields. Despite the isolated position of Masse's farm he had been suffering strange attacks of vandalism, with his crop of lavender bushes being torn up. Masse assumed that some local youths were to blame. He was a former fighter in the French Resistance and had decided that if he ever caught the culprits, he would teach them a stern lesson.

At about 6 am Masse stopped his tractor for a cigarette break. He heard a strange whistling noise that came from the other side of a small hillock. Jumping down from his tractor, Masse trotted around the hillock to investigate. He saw an egg-shaped object of gleaming silver metal mounted on six thin metal legs and a thick central support. It was about the size of a small van. Of more immediate interest to Masse were what he took to be two boys aged about 8 years old who were pulling at a nearby lavender plant. They had their backs to him.

Thinking he had finally caught the vandals who had been plaguing him, Masse crept towards them. When he was about 4.5 m (15 ft) from the 'boys' one of them seemed to hear him. The figure stood up and turned around, then quickly whipped out a small gun-shaped object and pointed it at Masse. The farmer stopped in alarm, but his alarm turned to fear when he realized that he had been paralysed and could not move.

While Masse watched, the second figure also stood up. Masse could now see that they were not boys at all, but bizarre entities. Each figure stood about 1.2 m (4 ft) tall with thin, slender bodies and limbs. They seemed to be wearing boots on their feet and there were shorts of some kind around the groin. Otherwise the figures were dressed in tightly fitting green overalls. The heads were oval, with a pointy chin and small ears. The eyes were large and oval in shape, slanting up to the side of the face. The mouths were thin slits without lips.

The figures made some strange grunting noises, which Masse took to be their speech. Strangely the mouths did not move, the sounds seeming to come from the throat area. The figure that had first seen Masse returned the gun-shaped object to a pouch

Based on the descriptions given by French lavender farmer Maurice Masse, this artwork shows the UFO and its crew that he encountered in 1965.

attached to the side of his belt. With a final look at the lavender, the two figures moved towards the egg-shaped object. A sliding door opened in the side of it and the two figures entered. They did not climb in, but seemed to float up from the ground.

The door slid shut and the object began to emit the whistling noise that had first attracted Masse's attention. There was a dome on top of the object. A window appeared in this and Masse saw one of the figures peering out of it at him. Then the window closed.

> ... his alarm turned to fear when he realized that he had been paralysed and could not move.

The object rose vertically into the air to a height of about 18 m (60 ft). The whistling noise then stopped and the craft disappeared off into the distance, towards the nearby village of Manosque.

An investigator scrapes up soil samples from the landing site of the UFO met by Masse. Nothing unusual was turned up by the detailed chemical analysis that followed.

Still Masse was paralysed. He now began to become seriously concerned, worried that he was permanently disabled and might die. It was not until some 20 minutes later that he regained the use of his body, though only slowly even then. Once he felt confident of moving, he began walking, then running for the village of Valensole. There he met the owner of the local cafe, who was just opening up for business, and blurted out his story. The cafe owner called the police.

Later that day Captain Valnet of the gendarmerie arrived, uncertain quite what had happened, but determined to get to the bottom of the trouble. The official investigation that followed collected evidence competently, including details of damage to the lavender and drawings of the marks left by the craft. It was noted that the plants around the landing site itself died over the following days. Those further from the spot where the craft had landed recovered, but those that had been under the craft did not. Masse himself was very tired for several days, sleeping much longer than was usual and being unable to undertake the more strenuous jobs around the farm.

Despite collecting a mass of evidence of various kinds, it is unsurprising that Valnet failed to reach a firm conclusion of any kind.

What makes the Valensole case so interesting is that the features reported by Masse were to crop up again and again in the years that followed, and remain typical in modern cases.

The aliens themselves fit the category that is now referred to as Tricksters. Their pallid skin, slanting eyes and pointed chins are typical, as are the rounded, bald heads and tightly fitting costumes. Their behaviour would also become typical. The interest of the aliens in plants, particularly crops, is something that is reported time and again. The strange floating movement of the aliens as they boarded their craft is another feature that was thought bizarre at the time, but has since become an almost routine element in reports of aliens.

Likewise the fact that they were neither hostile nor friendly to the human that saw them is a striking feature of this case. Quite clearly they did not want to be disturbed and took prompt action to disable Masse as soon as they realized that a human was nearby, but were otherwise unconcerned. They did not attack, and no lasting damage was done. Compared to some other types of aliens, which may attack on sight or flee, this behaviour is typical of the Tricksters.

The interest of the aliens in plants, particularly crops, is something that is reported time and again.

All of these features were reported by a French farmer who lived in an area about as rural as it is possible to get in western Europe, who had never taken any previous interest in UFOs, nor had ever read or seen any of the science fiction books or movies then being made. It would seem that the only explanation for this is that Masse did in fact have a real experience.

There is one final fact that is worth recording. Some years after the incident at Valensole, a French UFO investigator sent Masse a drawing of the UFO that had landed at Socorro in New Mexico, USA, as he thought it sounded similar to that seen by Masse. Masse's reaction was immediate and emphatic. 'That is what I saw,' he replied. 'You see, I was not dreaming and I was not mad.'

CHAPTER 6

Alien Contact

A very few people claim not only to have encountered the occupants of UFOs, but to have spoken with them. Sometimes the conversation seems to take place in the language of the human who encounters the beings, at other times it takes place telepathically or by some other unknown means.

One of the earliest such encounters, though it was not recognized as such at the time, took place outside Homan, Arkansas, in 1896. This encounter took place during the rash of sightings of mysterious airships over the USA that year. James Hooten, a railroad engineer, was out hunting in the woods near Homan when he heard a loud whooshing noise that reminded him of an air brake on a steam locomotive.

Pushing through the undergrowth, Hooten emerged into a clearing. Resting there was a large cigar-shaped craft with four paddle-like objects projecting out of the back and three large wheels along the side. A cabin hung down from underneath to touch the ground and at the back of this stood a

Hooten was out hunting in the Arkansas woods when he came across what he took to be human airmen, though modern investigators class his meeting as a close encounter of the third kind with aliens.

The German airship *Graf Zeppelin*. The UFOs seen over the USA in the 1890s were widely identified as being airships of this sort of design.

short man wearing a face mask. Hooten strolled up to the man and asked, 'Is this the airship?' The man seemed surprised to see Hooten, but replied: 'Yes, sir. Good day, sir.' There then followed a brief conversation in which the strange man told Hooten that the airship was powered by compressed air. Four other men then appeared and one said, 'All ready, sir.'

The strange men quickly ran inside the cabin. The wheels began to turn and the object emitted a whooshing or hissing sound. It then rose into the air, accelerated to high speed and shot out of sight.

A few months later, in April 1897, J. Lignon and his son were walking home across farmland one evening when they saw a light in a field. Going to investigate, they found a large, dark object with four men standing beside it. One of the men asked Lignon for water and Lignon pointed out a nearby stream. The men said that they were flying their airship from the Gulf of Mexico to Iowa and that the craft was powered by condensed electricity. Once the water was acquired, the men entered the craft by a door and then took off.

Going to investigate, they found a large, dark object with four men standing beside it.

Although Hooten and the Lignons interpreted their sightings to be of the mysterious airship seen over other parts of the USA, their reports included

Many witnesses have reported that the 'Greys' have large, dark eyes. The eyes seem to have an almost hypnotic effect on witnesses and some claim that messages are beamed into their brains via the eyes.

features that would later appear regularly in accounts of alien contacts. The vague and ultimately meaningless descriptions of motive force would prove to be particularly recurrent features.

Not everyone who encounters a UFO and its crew manages to converse quite as successfully as did Hooten and the Lignons. In October 1959, for example, Gideon Johannson had a power cut at his rural house in Mariannelund, Sweden. He and his son went outside to see if there were any obvious problems and saw an unusual flying object behaving in an erratic manner. Shaped like a bell with a large window on one side and glowing white, the UFO rocked from side to side, changed direction abruptly and crashed through the branches of a tree before coming to rest, hovering a short distance away.

Johannson went to investigate and was 3 m (10 ft) from the object when he suddenly noticed two humanoids through the window. He stopped abruptly when he realized that the ufonauts had also seen him. The figures were short with dome-shaped heads and pointed chins, but their main features were their eyes. These were large, dark and beautiful.

One of the beings used its eyes to transfer mental images to Johannson. These urged Johannson to stay where he was and showed the figures doing work on their craft, apparently repairing it. Johannson did his best to communicate back by thought or by waving, but the ufonauts studiously ignored him. After some minutes the UFO rose into the air, hovered and then emitted a bright flash after which it was gone.

When the power engineers came to repair the power cut they found that the lines leading to the Johannson house had been damaged about a mile away. Johannson believed that the UFO had inadvertently hit the cables, causing it to fly erratically and then come down to hover while the crew repaired the damage.

Equally one-sided was the conversation on 28 August 1963 at Sagrada Familia in Brazil. The Eustagio family was out in their backyard when a UFO came down from a cloudy sky. The UFO was round and white, descending with a humming noise and a rocking motion. It came to rest about 21 m (70 ft) in the air and two beams of bright light shot out from its underside to hit the ground.

One of the beings used its eyes to transfer mental images to Johannson.

Floating down in one of the beams of light was a humanoid figure. This being was an estimated

A UFO apparently hit overhead electricity cables near Mariannelund in Sweden in 1959, landing so that its crew could carry out repairs.

One of the more bizarre encounters came when a ufonaut asked Wisconsin farmer Joe Simonton for a jug of water. The alien gave Simonton four pancakes in return for the water. Simonton ate one of them, but did not care for its bland taste.

2.1 m (7 ft) tall and wore large boots with spikes coming out of the heels. An odd note was struck by the fact that it had only one eye, set centrally just above its tiny nose. The alien began blundering around the yard as if it were having trouble with the strength of Earth's gravity. After a few seconds it sat down heavily on a rock and looked around.

Only then did it seem to see the startled Eustagios who had been watching proceedings with concern while frozen still. The being then sent out a telepathic message telling the Eustagios that they had nothing to fear and that he, the alien, had come to explore the Earth on a peaceful mission.

While this message was being delivered, the alien got back to its feet and clumsily approached one of the Eustagio children. Even as he was telepathically assuring the Eustagios of his peaceful intent, the giant lunged at the child with flailing arms.

Things then moved quickly. The child fled while his elder brother picked up a heavy brick and ran screaming at the alien wielding the brick like a club. The alien staggered back, turned to face the assault and shot out a beam of light from its chest. The boy collapsed to the ground, temporarily paralysed. As the adults surged forward, the alien retreated back to the beams of light and floated back up to the UFO. The UFO hovered for a short while, then shot off at high speed.

Slightly more successful than Johannson or the Eustagios in conducting a non-verbal communication was farmer Joe Simonton who encountered a UFO near his home at Eagle River in Wisconsin, USA. The incident began as a fairly typical UFO close encounter when Simonton heard a rumbling sound and, on leaving his house to investigate, saw a silver-coloured oval craft about 9 m (30 ft) across and 3.7 m (12 ft) high flying over his house.

The ufonaut understood, picked up four pancake-like objects and handed them over to Simonton.

The object came down to land a short distance away. Suddenly a hatch opened on top of the object and three humanoid heads emerged. One of the figures climbed out. He was around 1.5 m (5 ft) tall and dressed in a tight-fitting outfit with a belt around the waist. The fabric was very dark blue, almost black, and seemed to be made of rubber or some such material. The creature's head was very human-looking with its straight, short black hair and swarthy complexion.

The alien had in his hand a large jug. He mimed as if drinking from the jug, from which Simonton concluded that his visitors were thirsty. He took the jug, filled it in his kitchen and brought it back.

As he handed over the jug, Simonton peeked into the craft through the hatch and saw one of the ufonauts apparently frying something in a pan over a sort of stove. Simonton pointed to the pan and mimed eating food. The ufonaut understood, picked up four pancake-like objects and handed them over to Simonton.

The alien then closed the hatch and the UFO flew off, leaving Simonton with the pancakes. The farmer phoned the local authorities, who passed the details on to the UFO researcher Dr J. Allen Hynek.

By the time Hynek and his team made contact with Simonton he had eaten one of the pancakes. It was, he said, fairly unpleasant, having a taste and consistency like that of cardboard. Hynek sent one of the other pancakes off for analysis. This showed it to be made of wheat bran, soya bean husks, buckwheat hulls and vegetable fats. In other words it was from Earth. Of course, that did not mean that the aliens had not cooked it, having developed a liking for such foods.

Sign language, if that is what it was, proved rather less successful during an encounter on 31 December 1974 at Vilvorde in Belgium. After sunset a local farmer went out into his yard to walk across to the outside lavatory located in a shed a few metres away. He had with him an electric torch, but was surprised to find that the whole area was suffused in a pale green light.

Looking round for the source of the light, the man saw a humanoid figure about 1.2 m (4 ft) tall walking across the far side of the yard. The figure had what looked like a rucksack on his back and was holding in front of itself a vacuum cleaner-like object which it was swinging back and forth as if using it to scan the ground.

One of the figures waved at Higgins, then beckoned him over. Rather apprehensively, Higgins approached.

The Belgian farmer switched on his torch and shone it at the intruder. The figure then stopped its scanning and turned to face the man. It held up its hands and moved them rapidly as if using sign language of some kind. After a final V sign with one hand the alien appeared to give up trying to make itself understood. It climbed over the wall and out of the yard.

A few seconds later a hissing sound came from the direction in which the figure had vanished. A disc-shaped UFO rose into sight over the wall, emitting a shower of sparks. It hovered for a short time, then flew off at high speed.

A more successful attempt at sign communication came on 23 July 1947 when José Higgins, a Brazilian surveyor, was working in a rural area near Pitanga. He had with him a team of local workmen to do the manual tasks. As they trampled over an open field the men heard a piercing whistle from overhead. They looked up to see a large disc-shaped flying object diving down towards them.

The workmen promptly threw down their equipment and fled, leaving Higgins alone with the disc. As the disc neared the ground a series of metallic poles came down to serve as landing legs. The object was, Higgins thought, about 45 m (150 ft) across and 4.5 m (15 ft) tall. There were windows set into the hull and through these Higgins could see two faces peering at him.

A flap fell down from the underside of the craft to form a ramp and out walked three humanoid figures. Each figure was about 2.1 m (7 ft) tall and dressed in a tightly fitting suit of some plastic-type material. Each carried a metallic backpack and wore sandals. They seemed to be generally human, although their legs were rather too long in proportion.

One of the figures waved at Higgins, then beckoned him over. Rather apprehensively, Higgins approached. The ufonaut then pointed into the hatch. Higgins peered in, seeing only a small empty room with a closed door apparently leading deeper in to the UFO. The humanoid pointed again and Higgins realized that he was being invited in, perhaps to go on a journey. He spoke to them in Portuguese, asking where they would take him. When the ufonauts showed no signs of understanding, Higgins resorted to making signs with his hands and exaggerated facial expressions.

Finally one of the beings seemed to understand. It pointed at the sun and said 'Alamo,' then drew a circle in the sand and repeated 'Alamo.' He then drew seven circles around the sun, with a dot on

Brazilian surveyor José Higgins hid in undergrowth after fleeing from aliens that he feared were trying to abduct him. From the safety of his hiding place, he was able to watch them at leisure.

each. He then pointed at the outermost dot and said 'Orque.' He pointed to Higgins, then to the UFO and finally to the seventh circle saying 'Orque' again.

Higgins had no intention of being taken to another planet, but did not fancy trying to outrun or outfight three alien beings, each of which was clearly taller and stronger than himself. He got out his wallet and showed them a photo of his wife. He again used signs to indicate that he wanted to get her so that she could join them on the journey to Orque. The aliens nodded.

Higgins then strolled off as casually as he could manage. Once on the far side of the field he dived into woodland, squirmed into a hiding place inside undergrowth, and then turned to watch events. The aliens spent the next half hour or so pottering about, looking at plants, throwing stones and seeming to waste time while awaiting the return of Higgins. Perhaps finally losing patience, the aliens then reboarded the UFO. The hatch slammed shut and, with the same whistling noise it had made earlier, the UFO took off and flew out of sight.

When the ufonauts show that they can speak the language of the witness, it makes communication much easier than with gestures and sign language. One such encounter took place in May 1940, though it did not become public until some years later.

Udo Wartena was working on his small gold claim in a remote valley near Townsend in Montana, USA, when he heard a loud rushing noise, rather like that of a turbine. He was familiar with the various army aircraft that flew overhead and assumed that this was one flying closer to the ground than was usual.

Turning to face the direction of the noise, Wartena saw what he at this stage took to be a military aircraft. It was hovering about 180 m (600 ft) away over a meadow through which meandered a stream

A hosepipe came down from the hovering UFO to dip into the stream.

that Wartena used as his water source when working at the claim. The craft was large: Wartena later estimated it to have been about 30 m (100 ft) long and shaped like a lens, with the central section about 10.6 m (35 ft) thick.

As Wartena watched he saw a section of the hull drop down to reveal a spiral staircase. Down this came a human-like figure dressed in a grey uniform with a cap. The new arrival began walking towards Wartena, so the prospector strolled forward. The figure waved at Wartena and as they got to within 3 m (10 ft) of each other, asked if it would be all right for him to take some water from the stream. Wartena said it was. The stranger then turned and waved back to his craft. A hosepipe came down from the hovering UFO to dip into the stream.

The man then invited Wartena on board his craft. Still thinking that he was dealing with the US military, Wartena agreed. He climbed the spiral stairs to find himself in a room about 3.5 m by 4.5 m (12 ft by 15 ft). The room was lit by a pale light, the source of which Wartena could not see. In the room was an older man seated on a plushly padded bench.

Wartena asked why the men wanted water from his small stream when there was a large lake not far away. The younger man replied that it was because the stream water was purer and contained no algae. Wartena had now realized that although the man spoke good American English, he was doing so slowly and stumbling over certain words as if this were not his native language. In May 1940 war was raging in Europe and in Asia and Wartena became suddenly suspicious of where these men were from, and so he asked them.

'We are from a different planet,' came the reply. 'It is a long way from here.' Wartena was surprised, to say the least, by this response but felt totally at

Montana prospector Udo Wartena let an alien extract water from a mountain stream. The alien spoke good English, though with an odd accent.

ease with the ufonauts. 'They seemed like very nice chaps,' he later reported.

The younger man then explained to Wartena how their craft worked. He said that there were two flywheels around the rim of the craft that span in opposite directions. This gave the craft an internal gravity negating that of Earth or any other planet. The craft gained its power by focussing the gravitational energy of whatever celestial body was the closest and then using this to skip over the light waves faster than the speed of light. They claimed to be able to store small amounts of the gravitational energy for emergency use.

Wartena then asked his new alien friends why they had come to Earth. He was told that they had been visiting Earth for some years to gather information, leave instructions and help out when they could. They then invited Wartena to come with them but he declined, saying that his friends and family relied on him and he could not just go wandering off without telling them.

At this the two men ushered Wartena off the UFO. They told him to get well away from the craft before it took off and not to discuss the event with anyone. In the event, Wartena did not move off far enough. The rushing noise began again and the rim of the UFO began to spin. It lifted vertically off the ground, hovered for a few seconds while wobbling slightly and then flew off. As it left Wartena collapsed. His muscles simply would not work and did not return to normal for more than an hour.

As instructed, Wartena stayed quiet about his meeting. In any case, he reasoned, nobody would believe him. It was only many years later, as reports of UFOs and flying saucers began to appear in the newspapers, that Wartena finally told people about his strange encounter.

In September 1955 Josef Wanderka was riding his moped through woods near Vienna, Austria, when he came upon a silvery, egg-shaped object resting in a clearing. A door in the side of the object was open and a ramp led down to the grassy glade. Gingerly, Wanderka entered to find himself in a large, featureless room where he was confronted by six humanoids.

A borg alien from the TV series *Star Trek.* Movie and TV shows often depict aliens as being basically human, dressed in one-piece suits, although the aliens reported by witnesses are far more varied.

The humanoids were rather taller than Wanderka and each had long blonde hair tied back in a bun or bunch. The ufonauts had faces that Wanderka likened to those of children. They were dressed in one-piece garments that reached from neck to wrists and toes. Taken aback, Wanderka stammered out his profuse apologies for having entered the craft, introduced himself and began backing out of the door.

One of the ufonauts then began talking in German with a slight accent. It urged Wanderka not to leave, assuring him that he was in no danger. The ufonaut said that they had come from the star system Cassiopeia and asked Wanderka how things were on Earth. This being the height of the cold war, with Austria caught between the Soviet and capitalist areas, Wanderka blurted out a quick summary of the then current international tensions. The ufonaut suggested that perhaps Wanderka could become Earth leader and so sort the problems out.

Wanderka was just wondering how to respond to this odd suggestion when the ufonaut changed the subject and demanded to know how the moped worked. After Wanderka had explained, he was ushered out of the UFO which then took off.

On the evening of 7 September 1957, James Cook came across a disc-shaped UFO on a small hill near Runcorn in northwest England. It was hovering a few feet above the ground and a ramp came down towards Cook. He then received a telepathic message inviting him on board, but warning him that he had to jump on to the ramp. If he touched the ramp and ground at the same, the disembodied voice said, he would be electrocuted.

Cook did as bidden and found himself inside an empty room lit by a soft glow emanating from the walls. A blue one-piece suit lay on the ground. He was instructed – again telepathically – to change into it. Having done so, he was told that the disc-shaped UFO was a short-range craft powered by electromagnetism used within Earth's atmosphere. He was now being taken to a larger craft, powered by ion drive, which was used for interstellar travel.

Having reached the mother ship, Cook was greeted by beings that seemed very similar to those encountered by Wanderka. They were tall, blonde and had the faces of children. Cook was told that these aliens came from the planet that they called Zomdic, but was given no indication of where it was. They said that they had no hostile intentions, but were worried by mankind's propensity for violence and war. They told him that humanity should work peacefully to sort out its problems.

They said that they had no hostile intentions, but were worried by mankind's propensity for violence and war.

Cook was then sent back to the smaller UFO and so transported back to Runcorn. When he got home, Cook found that 45 hours had passed, not the ten or so that he had thought.

Not all witnesses that converse with aliens are taken for space trips. On 24 April 1964 Gary Wilcox, a farmer from Newark in New York State, USA, was working a field when he saw a large object lying in another field beside a small wood. Thinking that the object was a disposable fuel tank from a military

The aliens that met US farmer Gary Wilcox were fascinated by his tractor and asked numerous questions about how it worked.

aircraft, or some such object, he drove over in his tractor to investigate.

The object was shaped rather like an elongated egg or stumpy torpedo about 6 m (20 ft) long and 1.2 m (4 ft) tall. It seemed to be made of a shiny, silver metal all of one piece without any signs of joints or rivets. Wilcox got down from his tractor and gave the object a hefty kick, causing it to give a metallic clang.

Suddenly two figures came out from underneath the object. They were about 1.2 m (4 ft) tall and dressed in white overalls that had a metallic sheen. One of them was carrying a bundle of alfalfa, apparently taken from Wilcox's field. Understandably, Wilcox began backing hurriedly away, but one alien called out to him, 'Don't be alarmed, we have spoken to people before.' Wilcox later thought that although he could understand clearly what was being said, the figure had not actually been speaking English but had been making strange bubbling and moaning noises.

The aliens then asked Wilcox if he could explain how his tractor worked. Wilcox did so, and then he was asked what he was doing with it. He explained that he had been spreading manure on a field, but the aliens did not seem to understand the concept of fertilizer, so Wilcox went on to explain that as well. The aliens said that their home planet was very rocky and unsuitable for growing Earth-type crops. They went on to ask Wilcox all sorts of questions about farming.

As the conversation seemed to be coming to an end, Wilcox asked if he could have a ride in their craft. The aliens refused, saying that the air that they breathed would be too thin for him. They said that they preferred to visit rural areas as the skies over cities were too polluted with fumes and gases that interfered with their craft's energy system.

The aliens said that their home planet was very rocky and unsuitable for growing Earth-type crops.

The aliens then disappeared back under the object. Wilcox presumed that there was some sort of door though he did not see one. The object then rose slowly and silently to a height of about 45 m (150 ft) before gliding off to the north until it was out of sight.

Also refused a ride was Heinrich Ivanovich, who encountered a UFO and its crew on the banks of the Kama River near Voronov, Russia, in 1969. He was riding home from work on his motorbike when he saw a man on the side of the road who raised his arm as if waving at him.

Ivanovich stopped and saw that the man was holding a hose that led from the river to a disc-shaped object resting on the ground some distance away. The man was wearing an all-in-one suit that seemed to be made of a grey metal fabric and he had on boots with abnormally thick soles. Otherwise he seemed to be completely human, speaking fluent Russian.

The man asked Ivanovich if he had an interest in astronomy, then moved on to talk about stars and constellations. After a while he pointed to the disc-shaped object and remarked that he had arrived in the craft, which was powered by electromagnetism.

Having apparently filled up his craft with water, the man told Ivanovich that he could not come for a journey. Instead he would have to stay on the road and watch the craft take off from a distance. The man then walked across the field, climbing into the object by way of a door that opened in its side. The object began to revolve, emitting a greenish light as it did so. Finally the UFO rose vertically skywards before simply disappearing into thin air.

One feature that was to become disturbingly common among those who talk to ufonauts was the missing time, a period during which the witnesses cannot recall what happened.

One early sufferer was police officer Herbert Schirmer after his encounter with a UFO on the night of 3 December 1967. Schirmer was driving just outside Ashland in Nebraska, USA, when he saw a number of bright lights strung along a side road. Driving over to investigate, he saw that the lights were fixed to a large oval object that had a rim or walkway running around its edge. The object was

Nebraskan policeman Herbert Schirmer writes out a report on his 1967 encounter with a UFO and its crew.

sitting on three legs and emanating a reddish light. A few seconds after Schirmer arrived the object took off with a wailing noise and flew away. Schirmer returned to his station to find that it was about 2 hours later than he thought. He logged the incident as a UFO sighting.

Some weeks later Schirmer was contacted by a UFO investigation group and persuaded to undergo hypnosis to see if he would then be able to recall any further details. Under hypnosis, he produced an account of what had happened during the missing 2 hours.

Schirmer recalled that as he stopped his patrol car in front of the grounded UFO, a humanoid figure had appeared beside it. This figure walked towards him carrying a box which directed a green beam of light on to the car. A second figure then appeared and walked up to Schirmer's open window. The figure leant in and touched Schirmer on the neck, causing a sudden but short stab of pain. The figure stepped back and Schirmer got out of the car.

'Are you the watchman of this town?' the figure asked.

'Yes, I am,' replied Schirmer.

'Watchman, come with me,' said the alien and turned to lead the way into the UFO.

Inside the UFO, Schirmer found himself in a room lined with large cylindrical objects that he likened to oil drums. Each of these had a black cable or pipe coming from it. From the ceiling hung a spinning half sphere that emitted various coloured lights.

'Watchman,' the alien said, 'this is our power source. It is reversible electrical magnetism. The reason we are here is to get electricity.' He went on to explain how the craft was extracting power from electric cables nearby.

The figure leant in and touched Schirmer on the neck, causing a sudden but short stab of pain.

Schirmer was able to get his first good look at the figure at this point. The alien was very like a human, but with a larger, slightly domed forehead and with eyes that had slit pupils like those of a cat.

The alien showed Schirmer some other rooms in

Schirmer was bathed in beams of coloured lights when he accompanied the ufonauts to their craft after they stopped his patrol car near Ashland, Nebraska.

Several UFO reports, like that of Tsiport Carmel, begin with the witness being woken up late at night by lights or sounds coming from outside the house.

the UFO, then took him back to the entrance door. 'Watchman, what you have seen and what you have heard you will not remember,' the alien told him. 'All you will remember is that you have seen something land and something take off.' Schirmer was then led back to his car and abandoned.

An encounter of a different kind took place at Shikmona Beach, Israel, on the night of 20 April 1993. Tsiport Carmel was woken up by a bright light streaming in through her bedroom window. Looking out she saw a disc-shaped object hovering close to the ground. It was emitting a bright light. A few feet from the UFO, Carmel saw a humanoid.

Hurrying out of the house, Carmel could now see that the ufonaut was well over 1.8 m (6 ft) tall and dressed in a one-piece, skin-tight blue suit that seemed to have a metallic glint. On his head he wore a helmet with a mesh-like mask.

'Why don't you take off your helmet?' asked Carmel.

'That's just the way it is,' the strange giant replied telepathically before returning to his UFO and leaving.

A very similar, possibly the same, ufonaut visited Hanna Somech in nearby Burgata a week or so later. He was first seen inspecting Somech's pick-up truck. When Somech's dog started barking, the tall man sent it flying through the air without touching it.

'What have you done to my dog?' demanded the startled Somech.

'Go away,' came the telepathic reply. 'I'm busy. I can crush you if I want.'

One of the most elaborate reports of a one-off conversation between a human and an alien came from Alicante, Spain, on 5 July 1978. Señor Pablo, a respected local businessman, was driving home through a rural district late that evening when he saw what he thought were the headlights of an oncoming car over the brow of a hill. He then noticed that the beams of light were of an intense orange colour, not the white or yellow to be expected of a car or lorry.

As he crested the hill, Pablo saw an orange disc-shaped object resting by the side of the road. He put his foot down on the accelerator, but the car engine spluttered and then cut out. The car rolled to a halt.

To Pablo's relief the object was now out of sight. He got out of the car intending to inspect the engine, but then heard a voice calling his name. Looking round he saw a man emerging from the darkness. The stranger was over 1.8 m (6 ft) tall and dressed in a tight one-piece outfit that covered him from neck to wrist and ankles. Although he looked basically human, the man had eyes that were larger than usual and slanted outwards.

Pablo was at first alarmed, but the man addressed him again by name and then opened a telepathic means of communication. Pablo became calmer at this point and later reported that he felt immediately convinced of the benign, friendly nature of the alien.

The alien explained that he had come from a planet far away in another solar system. His spaceship, he said, used a method of propulsion that acted outside the concepts of speed and distance using techniques beyond the reach of human science.

Pablo seems to have kept his head when confronted with this amazing individual. He certainly asked a string of questions that have troubled many of those who investigate the UFO and alien contact phenomena. Sadly the answers were not entirely illuminating.

The first of these questions was to ask why the alien appeared to be basically human when he came from an entirely different world. 'Of course,'

A schematic artwork of the solar system. Some ufonauts have claimed to have come from other planets, such as Venus or Saturn, although space probes have shown these planets to be incapable of supporting life.

was the reply. 'On our planet we are all humans just like you. There are certain anatomical differences but they are slight. However what makes us so different from earthlings is not the physical matter but the inequality of our evolutions, mental and then astral and spiritual.' The alien went on to explain that humans would one day make a similar astral, spiritual and mental evolution, but not for many years yet.

Pablo next asked if there were different sorts of aliens. 'Of course,' the visitor said. 'Some come to Earth in peace, others to observe you, but some are not your friends.' He went on to say that there were around 120,000 forms of intelligent life in the galaxy, not all of which had achieved interstellar flight. Venus, Mars and Earth's moon had once supported life, but not any longer. There was, he said, an advanced civilization on Jupiter's moon Ganymede. There was another on Neptune, but it was of a spiritual type that humans would not be able to detect.

The alien went on to explain that humans would one day make a similar astral, spiritual and mental evolution ...

Pablo then posed the key question as to why the alien had chosen to appear to him on a remote country road rather than to an important politician in the middle of a city. The alien seemed to find the question amusing. He explained that politicians lacked the spiritual strength to understand the message he had to impart and so the aliens sought out those with astral talent, such as Pablo.

Pablo then turned to religious questions. The alien asserted that there really was a God, but that the supreme deity was not quite as humans imagined Him to be. The key thing, he said, was to love God. He added that Jesus Christ had been misunderstood during his time on Earth, but that again the message of love was paramount.

The alien then moved on to warn Pablo that humanity was on the edge of catastrophe. Sadly he did not elaborate on what form this disaster would take or how to avoid it.

Instead he reverted to speaking about his home planet. This place had been, he said, made uninhabitable by a terrible ice age that had covered the entire planet with glaciers. It was this, he said, that had driven the evolution of his ancestors away from the mere material to the spiritual so that they could now survive on cosmic energy which they called Abuchal.

The alien then gave his name as being Naazra-Abuc before turning to walk off into the darkness. A few minutes later, Pablo saw the orange disc lift into the air and fly off.

The story related by Pablo is interesting as it links the majority of reports of conversations with aliens with the more complex stories told by contactees who claim to have had dozens of conversations with aliens. In general those who claim to meet and talk to aliens have only one encounter.

The alien encountered by Spanish businessman Señor Pablo said that Jesus Christ had been misunderstood during his time on Earth.

The conversation usually includes references to the motive power of the UFOs, which while sounding impressively technical are vague to the point of meaninglessness – rather like *Star Trek*'s warp drive or *Dr Who*'s sonic screwdriver. The aliens usually also pass on some vaguely worded warnings about humanity being in terrible danger of self-inflicted disaster, but again details are scarce and unhelpful.

These features are, indeed, present in Pablo's encounter, but he also went on to discuss religious and spiritual matters which links his conversation to those of the contactees.

The first of the witnesses to claim to have become a regular contactee for aliens was George Adamski (See Encounter Casebook No. 8 on page 154). Adamski's 1952 encounters were widely reported and created a sensation, though his later claims were widely discredited.

> **The aliens usually also pass on some vaguely worded warnings about humanity being in terrible danger ...**

Beginning before Adamski's claims became public was the rather less publicized experience of Orfeo Angelucci, a worker at Boeing's aircraft works at Burbank, California. Angelucci claimed that on 24 May 1952 he had been driving home from his regular night shift when he spotted a red UFO and stopped his car to get a better look. The UFO then sent out two much smaller green globes which came down to hover beside Angelucci's car. In response to a telepathic voice, Angelucci got out of the car to be confronted by a man and a woman of startling beauty who exuded an aura of wisdom, peace and nobility.

The two beautiful people proceeded to lecture Angelucci on the dangers facing humanity and how an unspecified race of aliens was seeking to find ways to save mankind from itself. They then vanished and the green globes flew off to rejoin the red UFO. Angelucci kept quiet about the bizarre encounter.

Two months later he again encountered a UFO. This time a door opened and Angelucci climbed in. The door slammed shut and the UFO took off at high speed. Within minutes, Angelucci claimed, he was gazing back at the planet Earth from outer space. The UFO seemed to be remote controlled as no aliens were on board, but a voice again lectured him on the desperate plight of mankind. Angelucci suddenly found himself bathed in a bright white light which caused him to undergo a spiritual reawakening.

In later contacts with various aliens, Angelucci was told that unless mankind improved it would be wiped out in 1986 by some unspecified disaster. Angelucci published his claims in a book entitled *The Secret of the Saucers* in 1955. He wrote several more books until the last in the 1960s, which was not generally taken very seriously. He died in 1996. Although he is largely ignored today, Angelucci did leave one lasting legacy when he coined the phrase 'New Age' to describe the epoch of human history that the aliens would usher in.

The UFO seemed to be remote controlled as no aliens were on board, but a voice again lectured him on the desperate plight of mankind.

American aircraft worker Orfeo Angelucci claimed to have met two startlingly beautiful aliens in 1952.

Another contactee to come forward with astonishing claims was London taxi driver George King. King had studied yoga and other eastern religious and philosophical systems for a number of years when, in May 1954, he was relaxing alone in his London flat. According to King's later account, he was suddenly startled by a disembodied voice which boomed out to declare: 'Prepare yourself. You are to become the voice of the Interplanetary Parliament.' Eight days later an Indian holy man materialized in King's flat to teach him the advanced psychical methods needed to contact the cosmic masters of the Interplanetary Parliament.

King claimed that, having mastered these techniques, he duly contacted a being from the planet Venus named Aetherius. This highly spiritual being acted as King's link to the cosmic masters

London taxi driver George King claimed to have been visited by an ancient Indian holy man in his flat, as well as meeting a ufonaut from another planet.

and taught King the true history of humanity. This secret story began millions of years ago on the planet Maldek, located between Mars and Jupiter. There humanity produced advanced civilizations comparable to those that flourished on Venus, Mars, Jupiter, Saturn and Uranus. Unfortunately the humans abandoned spiritual energy for mechanical energy, blasting Maldek to pieces when a complex nuclear experiment went badly wrong. Maldek was reduced to a collection of asteroids, which orbit the sun between Mars and Jupiter.

King's version of mankind's prehistory continued by claiming that some humans had landed on Earth to found a civilization on the continent of Lemuria in the Pacific. This civilization also overreached itself and destroyed Lemuria, but not before founding a colony on a large island in the Atlantic. This was Atlantis which, likewise, destroyed itself with nuclear experiments. Mankind was thus reduced to isolated bands of survivors who turned to stone age subsistence to survive. Humanity had only just returned to some form of civilization, King claimed, and was now in the 1950s risking it all again with frighteningly dangerous experiments with atomic and nuclear technology.

The cosmic masters had therefore decided to contact mankind to persuade them to abandon technology and instead take up the spiritual path chosen by the more enlightened civilizations of the solar system.

King then established the Aetherius Society, publishing a regular journal named *Cosmic Voice,* and began spreading his message through public lectures and media appearances. In 1959 he moved to the USA, settling in Santa Barbara, California. He died in 1997, but the Aetherius Society continued to flourish.

The society survived the discovery by science of the dead, sterile nature of the other planets in the solar system as they believed the beings with which they were in touch were spiritual in nature, not material. The aims of the Aetherius Society remain to spread the teachings of the cosmic masters and to create favourable conditions for closer contact, and a meeting, between the cosmic masters and members of the spiritual hierarchy of Earth.

Meanwhile, they claim that their chief aim is to perform advanced metaphysical missions in co-operation with the cosmic masters in order to benefit the mother Earth as an entity as well as mankind as a whole. To this end they charge up spiritual energy radiators by concentrated prayer. The radiators then beam the beneficial forces towards trouble spots in the world. Among the more recent successes claimed by the Aetherius Society was the relief of the distress felt by flood victims in Jakarta in February 2007. The dousing of raging forest fires in Kosciusko National Park in New South Wales in January 2007 was claimed as another success that followed the alleged release of spiritual power from an Aetherius Society radiator.

There can be little doubt that the followers of the Aetherius Society sincerely believe what they say, and that they are attempting to help their fellow humans by channelling spiritual energy to the world's trouble spots.

It is just as certain, however, that the claims of the various contactees are not considered to be part of the UFO phenomenon by most of those who study and research UFOs and alien contacts.

After standing around in ominous silence for some time the men began communicating with Bender by telepathic means.

Meanwhile, a different but related type of contact with UFO-related entities was developing. These encounters have become known as 'men in black', or MiBs, though not all of them involve meeting with men dressed in black.

One of the earliest MiBs took place in 1954 when UFO investigator Albert Bender was at home in Connecticut, USA. At the time Bender was head of the International Flying Saucer Bureau and published a regular journal entitled *Space Review*. Bender was visited by three men dressed in smart black business suits and wearing black, homburg-style hats. They kept the hats pulled down so that the brims partially hid their faces. One of them was carrying a copy of *Space Review*. After standing around in ominous silence for some time the men began communicating with Bender by telepathic means. They told him that he had to stop his UFO investigations at once. They issued threats and made claims that terrified Bender.

Next day Bender ceased publication of *Space Review* and resigned from the International Flying

Saucer Bureau. He later moved to the west coast of the USA, cut off all contact with his friends and insisted on having an unlisted phone number.

Thereafter several UFO witnesses, and some investigators, began reporting threatening visits from men dressed in black, or in official uniforms of one kind or another.

In November 1961 Paul Miller saw a landed UFO and two humanoids in North Dakota. On the following day he was visited in his office by three men in black suits who claimed to be from the government. They asked him detailed questions about the UFO and demanded to be shown the clothes which Miller had been wearing at the time. So frightened was Miller by the threatening behaviour of the men that he drove them to his house and showed them the clothes. After inspecting them, the men left.

In 1967 Robert Richardson sighted a UFO near Toledo, Ohio. He saw it land and later picked up a piece of metal from the landing site which he passed to a scientist for analysis. Three days later two men pulled up outside Richardson's house in a 14-year-old black Cadillac. They asked him a few innocuous questions about the UFO, then left.

A week later two different men arrived in a different car. They wore black suits and spoke with foreign accents. They acted in a threatening fashion and sought to bully Richardson into accepting that the UFO sighting had been a mistake or dream. Then one of the men asked for the piece of metal. When Richardson said he no longer had it, the man turned angry. 'If you want your wife to stay as pretty as she is, you had better get that metal back,' the intruder declared. Then both men left the house hurriedly.

A sketch by Albert Bender of one of the mysterious men in black who visited him in 1954 to utter threats.

It subsequently turned out that both cars had fake number plates and that the metal was a fairly nondescript iron alloy. The men were never seen again and their threats were not carried out.

The threats made to Mexican UFO witness Carlos de los Santos also turned out to be empty. The young pilot had seen a UFO when flying over southern Mexico in 1975. He was driving through Mexico City on his way to a television interview when two black cars overtook him and forced his car to the side of the road. Four large men dressed

A scene from the 1997 movie *Men in Black* starring Tommy Lee Jones and Will Smith. The film showed fictional government agents who suppress sightings of aliens while dressed in black.

in black suits and wearing black hats got out of one car. One of them peered in through Santos' open car window. He glared at him for a long time without blinking, then declared, 'Look boy, if you value your life don't talk any more about this sighting of yours.' Then the men got back in the car and drove off. Santos did continue to talk, but nothing happened to him.

Canadian UFO witness Carmen Cuneo was approached in 1976 by a man in dark clothes who told him to stop repeating his story or risk suffering serious, but unspecified injury. Again the threats were not carried out.

One case of men in black who did not utter threats came to light in 1964. On 24 May Jim Templeton took his family for a picnic on Burgh Marshes near his home in Carlisle, England. The family had a lovely time and noticed nothing unusual, other than the fact that cattle in a nearby field seemed rather nervous and prone to galloping about for no apparent reason.

However, when Templeton got his photos developed he was in for a shock. One of the photos he had taken of his daughter showed a strange figure in the background. The figure looked like a man dressed in a tight-fitting white suit with a visored helmet on his head and one arm on his hip. Templeton sent the photos and negatives back to the camera company for an explanation, but they could find nothing to indicate that the photo showed anything other than what was there at the time – yet Templeton and his family were adamant that no man in a space suit had been present. The photos were later given to the police who likewise failed to explain them.

A few days later, Templeton got a phone call from a man who said that he worked for the government and asked if he could come to interview him about the incident. Templeton agreed and after a few minutes two men knocked on his door. They

The unexplained photo taken by Jim Templeton in 1964 which clearly shows a strange figure behind his daughter. The photo led to a classic men in black encounter.

showed him apparently genuine papers identifying themselves as working for the military police.

After a few minutes' conversation about the incident, the two men offered to drive Templeton out to the site of the picnic. When they got to Burgh Marsh, the two 'military policemen' turned nasty. They began demanding that Templeton admit he had faked the photo or that an innocent passer-by had been present. Templeton refused, whereupon the two men left abruptly and drove off. Templeton had to make his own way home, a journey of about 8 km (5 miles).

The British military later insisted that the two men did not work for them and that they had no interest in the matter.

Visits by men in black have become less common in recent years, and opinion is divided among UFO investigators as to who they might be. Some believe that they are government agents seeking to hush up particular UFO sightings. Others think that they are aliens posing as humans for a similar purpose. A third theory holds that the MiB are hallucinations prompted by the trauma of encountering a UFO.

The truth is that nobody is entirely certain who the men in black might be. It is clear, however, that so far as is known none of their threats have ever been carried out and nobody has ever suffered at their hands.

The same cannot be said of the aliens themselves. Some people claim to have been abducted by aliens and to have suffered some very real injury as a result.

Encounter Casebook No. 8

TYPE>>**Contactee** DATE>>**Various dates, 1952** PLACE>>**Mojave Desert, California, USA** WITNESS>>**George Adamski**

Adamski shot to international fame with the publication of the book *Flying Saucers Have Landed.* The book was written by the British writer Desmond Leslie, but was based for the most part on claims made by Adamski about his 1952 encounter.

In this original version of his experiences, Adamski said that he and some friends had been enjoying a picnic in the Mojave Desert on 20 November 1952 when they saw a large cigar-shaped UFO pass overhead, chased by some military jets. As the UFO fled, a smaller disc-shaped UFO seemed to detach itself and came down to land a mile or so away.

While his friends waited, Adamski set off to investigate the landed flying saucer. As he approached, he was met by a humanoid who was dressed in brown overalls. The ufonaut could speak no English but made himself understood by hand signals and telepathy.

The being explained that he had come from the planet Venus as a messenger to Earth. The Venusians were a highly advanced civilization, but were deeply worried by the wrong turn that humanity had taken technologically by taking up nuclear power for both warlike and power generating purposes. The Venusian said that not only was humanity at risk, but so were other interplanetary races as the radiation was leaking out

A photo taken by George Adamski which he claimed showed a Venusian mother ship with four smaller saucers emerging from it.

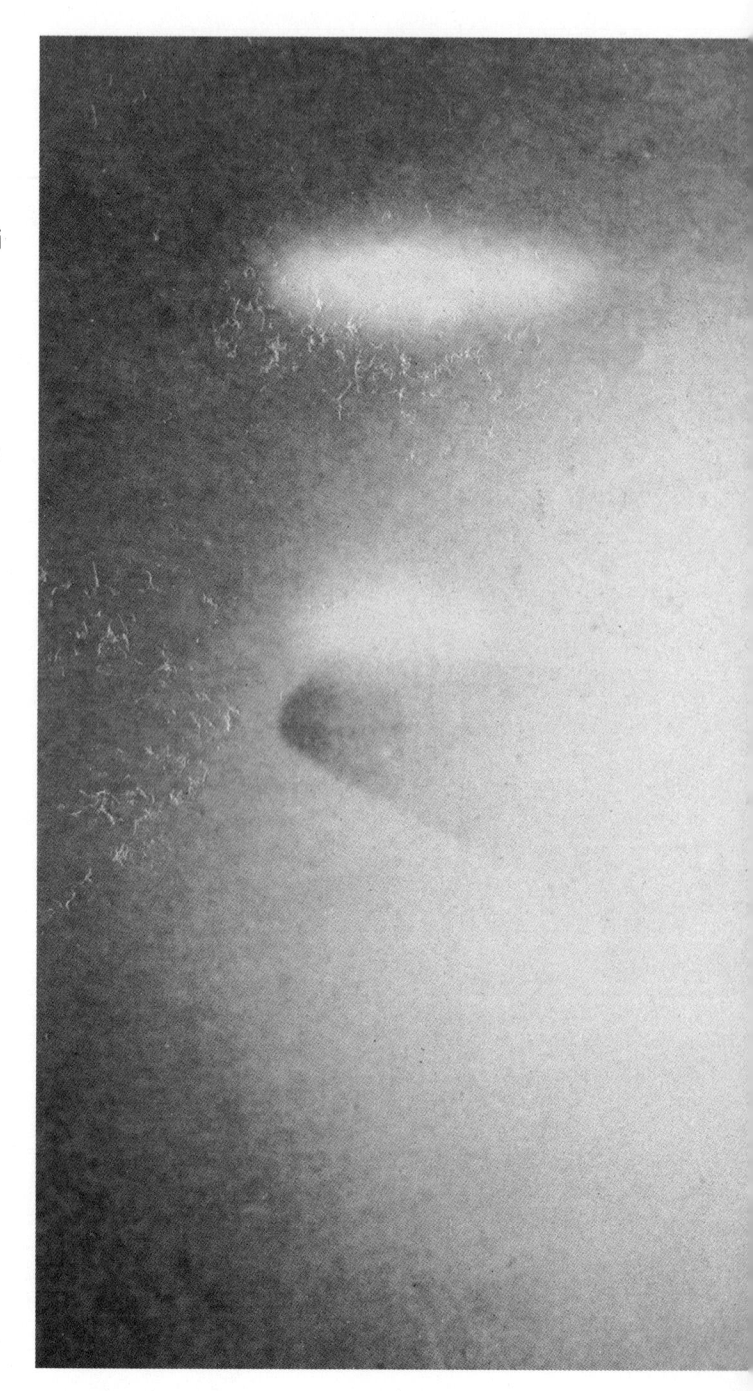

into space from Earth. The Venusian explained that he and his race wished to correct the path of humanity by peaceful means, but warned that other races – from such planets as Jupiter, Saturn and interstellar planets – were not so well disposed and might turn to force if mankind did not change its path voluntarily.

> The ufonaut could speak no English but made himself understood by hand signals and telepathy.

One of the clearest photos produced by George Adamski to back up his claims of alien contact was this one taken in 1952. He said it showed a scout craft of the type in which he had travelled himself.

The Venusian then asked to borrow a roll of photographic film that Adamski had on him. He agreed and handed it over, whereupon the alien left.

Adamski hurried back to his friends with his news. They agreed to sign a statement confirming their viewing of the original UFO and the smaller saucer.

By 1960 his claims had become outlandish and quite obviously he had no evidence at all to support them ...

On 13 December, Adamski returned to the desert to retrieve his film. He was visited by a flying saucer, the film dropping from a porthole. When developed it showed a variety of spacecraft.

These original claims created a sensation. Some believed Adamski, others condemned him as a charlatan. His background as part-time philosopher, self-appointed professor, one-time alcohol bootlegger during prohibition and full time burger stand salesman did not inspire confidence.

In 1955 two of the friends who had been on the fateful picnic came forward to say that in fact they had seen no UFO. They had signed the statement prepared by Adamski to help him sell the story for cash, but now that the story was becoming so widely believed they felt that they had to retract.

In the years that followed, Adamski claimed to have been visited by the Venusians on several more occasions. He also claimed to have been taken to the moon and to Venus, where he found lush forested valleys. By 1960 his claims had become outlandish and quite obviously he had no evidence at all to support them, apart from the photos of spacecraft that photographic experts proclaimed to be clever fakes made by photographing detailed models in carefully controlled artificial light.

Adamski died in 1965. Soon afterwards space probes would reveal that the planets he claimed to have visited were quite unlike his descriptions and that places Adamski claimed to be the home of advanced civilizations were dead, sterile worlds.

Encounter Casebook No. 9

TYPE»**Man in black** DATE»**11 September 1976** PLACE»**Orchard Beach, Maine, USA** WITNESS»**Herbert Hopkins**

In the mid 1970s Dr Herbert Hopkins was using his hypnotic skills to assist UFO researchers to probe the memories of UFO witnesses. People under hypnosis are often able to recall details that have been lost from their conscious minds. But the direct intervention and threats of a mysterious man in black caused Hopkins to drop the case he was working on.

In September 1976, Hopkins was helping with the witness David Stephens who had seen a UFO in bizarre circumstances in the previous October. On 11 September, Hopkins' wife and children had gone out for the evening when he received a phone call from a man claiming to be vice president of the New Jersey UFO Research Organization. The man said he was in the area and asked if he could drop by to discuss the Stephens case. Hopkins agreed.

Just moments later the man arrived. He was dressed in an immaculate black suit with sharply creased trousers, black hat, black shoes, black tie, grey gloves and white shirt. Hopkins let the man in and asked him to sit down. The man removed his hat to reveal that he was totally bald and that his head and face had a pale, whitish colour.

> He was dressed in an immaculate black suit with sharply creased trousers, black hat, black shoes, black tie, grey gloves and white shirt.

A swinging watch is a traditional method of hypnosis. Many UFO witnesses have remembered clear details under hypnosis.

The two men chatted for some time about UFOs in general and the Stephens case in particular. Hopkins noticed that his guest spoke in a curiously flat, emotionless monotone. Odd as this was, even stranger was the fact that the man was wearing lipstick that came off on his glove when he brushed his lips with his fingers.

Suddenly the man asked Hopkins for a coin. Hopkins handed it over and watched in great surprise as it vanished from the man's open palm. The man then explained that he could make a heart

The man then began to slur his words. He stood up abruptly and announced: 'My energy is running low. I must go now.'

vanish from within a human body, just as he had made the coin disappear. He then ordered Hopkins to stop working on the Stephens case and to destroy all his files and information. By now thoroughly terrified, Hopkins agreed.

The man then began to slur his words. He stood up abruptly and announced: 'My energy is running low. I must go now.' Walking stiffly and with apparent difficulty the man left the house and went out to the driveway. Hopkins saw a bright bluish light that he took to be car headlights.

When Hopkins' family arrived home he told his wife of his strange visitor. Together they inspected the drive and found odd markings unlike those that would have been left by a car or motorbike. Hopkins quickly discovered that there was no such thing as the New Jersey UFO Research Organization.

Deeply worried about who, or what, his visitor might have been, Hopkins dropped the Stephens case and destroyed his files.

A man in black who visited Herbert Hopkins made a coin disappear in eerie and ominous circumstances.

CHAPTER 7

Alien Abductions

No other facet of the entire alien encounters enigma has attracted more attention in recent years than that of abduction. The nightmarish quality of the reports makes these stories perhaps the most dramatic and intense alien encounters yet reported. Although the experiences vary a great deal, they nearly all fit into a definite pattern that has potentially disturbing implications.

One unusual case that forms a sort of bridge between the encounters that we have studied so far and the alien abduction proper is that of Jan Wolski, a 71-year-old Polish farmer. One morning in the summer of 1978 Wolski was driving his horse-drawn cart through a wood near his home when he saw ahead of him two small humanoids.

The beings were about 1.2 m (4 ft) tall and very slender in build. They were dressed in tight-fitting one-piece outfits of a silver-grey colour. Their heads were larger than a human's would have been in proportion to their bodies, and had large, almond-shaped eyes of a very dark colour. Their ears and noses were very small and their mouths little more than slits. Bizarrely, they were bouncing along as if their shoes contained hidden springs.

> When Wolski came closer, the two creatures noticed him and bounded over to sit up on the cart.

A farmer and his cart in the vast forests of Poland. It was in these remote woods that Jan Wolski had a bizarre encounter.

When Wolski came closer, the two creatures noticed him and bounded over to sit up on the cart. One of them pointed forward along the lane, seemingly indicating that Wolski should continue. The two beings then chatted to each other in a language that Wolski did not understand. He did, however, get the very strong impression that the beings were in a jovial and happy mood. After a few minutes the cart entered a clearing and the two beings jumped down.

They began bounding off towards a UFO that was hovering some feet above the clearing. The object was white and shaped rather like a house, with a

Wolski's encounter began when he willingly gave a lift on his cart to two short humanoids that he encountered in the woods. They led him to their UFO.

pitched roof like that of a barn. At each corner were cylindrical objects from which projected vertical black poles topped by spinning spiral objects. A loud and intense humming sound was filling the air. As Wolski watched, a black box-like object began descending from the UFO on four cables.

One of the humanoids then turned to Wolski and beckoned him over. Since he had felt no impression of hostile intent, Wolski climbed down from his cart and walked across the clearing to the UFO. The being then gestured for him to enter the box, which he did. The box then ascended into the UFO.

Stepping out of the box, Wolski found himself inside a very gloomy room. There were two large tubes extending from one side of the UFO interior to the other, and a number of rods in the floor that Wolski took to be controls. Also on the floor were a dozen or so birds, apparently paralysed or dead.

One of the ufonauts indicated by hand signals that Wolski should undress, which he did. The aliens then studied Wolski visually and passed what seemed to be scanners over his body. He was then instructed to get dressed again and was shown back to the box-like lift apparatus. The lift descended with Wolski and the two ufonauts in it.

Telling them to go through their drill of defending a campsite, Valdes put out the fire and got his men into defensive positions.

As Wolski exited the lift he thought that he should bid his odd companions a polite goodbye, so he turned to bow and doff his cap. The aliens bowed in return, and then the lift went up back into the UFO. The UFO rose slowly into the air before departing at high speed.

The Wolski encounter is interesting. In many ways it resembles an alien contact of the type we have seen already. He meets a UFO and its crew, is given a tour of the UFO and then allowed to leave. Like some Trickster ufonauts, the beings Wolski encountered had an interest in local wildlife, as shown by the birds on the floor. However, the encounter also has elements of an abduction experience about it. Wolski was undressed and subjected to a physical examination, albeit of a relatively brief and non-invasive nature.

Perhaps crucially, the beings Wolski met match very closely the description of the beings known to researchers as Greys. It is these Greys that have been involved in the vast majority of abductions.

Also on the borderline between an alien encounter and an alien abduction was the incident that took place just outside Putre in Chile on 25 April 1977. Six conscripted soldiers were camping out on a remote mountainside under the guidance of a regular corporal, Armando Valdes, as part of their basic training.

After pitching camp, lighting a camp fire and other outdoor basics, the squad settled down for the night. Valdes apportioned guard duty to his conscripts, with orders to wake him if anything unusual happened. Some time after 4 am conscript Pedro Rosales was on sentry duty when he spotted some unusual purple lights floating down from the sky towards the campsite.

Rosales awoke Corporal Valdes to see the purple object. Uncertain what it was, Valdes thought this an ideal opportunity to conduct an exercise with his men. Telling them to go through their drill of defending a campsite, Valdes put out the fire and got his men into defensive positions. The glowing lights had by this time landed about 180 m (600 ft) away up the slope among some undergrowth.

Valdes told his men to keep him covered, then picked up his rifle and walked up the slope towards

Night-time manoeuvres can be unnerving for military recruits, but one patrol in Chile in 1977 led to terror and mystery.

the purple object. He was about halfway to the object when there was a flash of light which half-blinded the conscripts. The hillside was then plunged into darkness, made worse by the fact that the men's night vision had been ruined by the flash. Both the purple object and Corporal Valdes had vanished.

The conscripts were deeply unnerved by what had just happened. Their unease about the situation was made worse by the fact that they no longer had their corporal to tell them what to do. As the men debated in hushed tones what they should do, Valdes suddenly reappeared – some 20 minutes after he had vanished.

Valdes came staggering down the slope as if in a daze or slight trance. He stopped in front of the conscripts, swaying on his feet. Then he began to mutter. 'You do not know who we are,' he said. 'You do not know where we are from. But I tell you this – we shall return.' He then pitched forward on to his face in a dead faint.

The conscripts hurriedly relit their fire, made Valdes comfortable in a camp bed and stared warily into the darkness, their fingers on the triggers of their rifles. When dawn came they relaxed slightly and decided to carry Valdes down the mountain to Putre. The conscripts noticed two odd things about Valdes. Firstly, he had stubble on his chin as if he had not shaved for several days, when he had been clean shaven the day before. Secondly, his digital

Charlie Hickson and Calvin Parker were fishing in Mississippi when a craft appeared and they were abducted by three strange-looking aliens.

watch had stopped dead at 4.30 am on 30 April, which was five days into the future.

Arriving in Putre, the soldiers headed for the house of the local schoolteacher. He called the army, who arrived promptly to take Valdes off to military hospital. Valdes recovered quickly, but was never able to recall what happened to him on that lonely mountainside in Chile.

Another early alien abduction that did not quite fit the pattern which would later become established was the Pascagoula case, which was later to become controversial among UFO researchers.

On the evening of 11 October 1973, two shipyard workers ran into the police station at Pascagoula in Mississippi, USA, claiming that they had been abducted by aliens. Sheriff Fred Diamond did not at first believe a word of the story, but realizing that the men were clearly terrified, he agreed to take a statement. Both men, Charlie Hickson and Calvin Parker, told near identical tales.

The two men had been fishing for catfish from a pier in an isolated area off Highway 90. They had heard a loud zipping noise behind them and turned to see an oval-shaped object hovering about 0.9 m (3 ft) off the ground. The object was glowing with a pale blue light and seemed to have windows. A buzzing noise filled the air.

> Both Parker and Hickson said that they thought the beings were robots rather than living aliens ...

As the men wondered what they had encountered, a door opened in the side of the craft and three humanoids came floating out. The figures were about 1.5 m (5 ft) tall and glowed with a soft white light. Their heads were high and domed with large, black eyes and slits for mouths. Their ears were small, conical and stuck out sideways from their heads. Their arms ended in claw-like hands with simple pincers instead of fingers. Both Parker and Hickson said that they thought the beings were robots rather than living aliens, though they had no real way of deciding this.

One of the aliens grabbed Hickson by the shoulder, causing him to feel a strange stinging or tingling sensation. He then drifted into a trance-like state in which he was only vaguely aware of what was happening. Parker passed out completely at this point and was unable to remember anything much.

According to Hickson, both men were levitated by the aliens and floated into the UFO. Inside the craft, the two men were subjected to what appeared to be medical tests of some kind. Hickson thought that they were also photographed, but the only really clear recollection that he had was of lying on a table while a gigantic scanning device shaped like an enormous eye was passed repeatedly over him.

The aliens then floated the two men back out to the riverbank and dumped them on the ground next to their abandoned fishing tackle. One of the beings then sent a message to Hickson which he assumed was arriving telepathically as he did not actually hear a voice. It said: 'We are peaceful. We mean you no harm,' repeated several times. Parker came round at this point, just in time to see the aliens leave and the UFO fly off.

By the time he had finished taking down the statement, Sheriff Diamond was nonplussed. The story sounded too bizarre to be true. To be sure, Diamond then left the men alone in the interview room, but continued to listen to what they said. He was half expecting the men to start gloating about how their hoax was fooling the police, but instead Hickson and Parker continued to talk about their encounter in worried tones. Parker said that he wanted to see a doctor, while Hickson said that he was of the view that nobody would believe them.

Now confident that the two men were telling him what they believed to be true, Sheriff Diamond arranged for them to be taken for a medical examination at a nearby USAF base. The doctor found no signs of any injuries nor of radiation, though both men did have a small cut similar to that used when taking blood samples.

A couple of days later Parker had a nervous breakdown and was admitted to hospital. Hickson, however, remained able to answer questions. He took a lie detector test which he passed and then underwent hypnotic regression to see if he could recall any further details about the encounter, but little more was revealed.

The case appeared to be a dramatic and impressive account of abduction by aliens. However, later investigations found that the lie detector test had not been carried out according to standard procedure so the results could not be relied upon. Moreover, the man who was on duty that night at a toll bridge just 270 m (900 ft) from the site of the encounter was questioned and said he had seen nothing all evening.

Some researchers chose to reserve judgement on the case, but most have accepted that Parker and Hickson really did encounter some strange aliens on the banks of the river that evening and that they were abducted. The two men later turned down a movie contract that would have brought them a considerable sum of money, saying that they just wanted to forget the whole affair.

A rather similar abduction took place on 22 September 1971 near the town of Itaperuna in Brazil. A mechanic, Paulo Silveira, was driving home from visiting friends that evening when he became

Calvin Parker gives evidence about his abduction by aliens. He later shunned publicity and tried to forget the whole incident.

aware that his car was being followed by a low-flying aircraft that seemed to dip in and out of sight. Spotting a payphone, he pulled over and called the police. They logged the call, but said there was nothing they could do.

Silveira drove on and, turning a corner, saw the strange aircraft sitting on the road. The car engine promptly died and the vehicle coasted to a halt. The object in front of Silveira was about the size of a large car or small van. There were small windows and an open door. Beside the door stood two humanoids about 60 cm (2 ft) tall.

These creatures had large, flat-topped heads, small eyes and pale skins. They were dressed in tight overalls that seemed to glow faintly blue. Each wore an open-faced helmet with a spike on top. They carried oddly shaped objects in their hands that pulsed with red and blue light.

Silveira felt all energy drain from him. The aliens approached, opened the car doors and began leading him towards their UFO. As he reached the door, Silveira seemed to snap out of his trance-like state. He tried to break free and run off, but the beings grabbed his arms with claw-like hands that scratched his flesh and held on with enormous strength. After a few seconds the lethargy returned and Silveira gave up fighting.

Inside the UFO, Silveira was met by five more ufonauts. The creatures began prodding and poking the human. There then came a loud noise like a passing railway train and Silveira passed out. When he came to he was being carried outside the UFO and laid gently down on the grassy verge beside his car. The aliens left him and hurried back to their craft. The UFO then rose slowly into the sky before emitting a brilliant flash of light and shooting off at high speed.

Silveira tried to break free and run off, but the beings grabbed his arms with claw-like hands that scratched his flesh.

Alone again, Silveira was unable to move. The overwhelming lethargy meant that he could barely sit up. A few minutes later a car pulled over. The driver, Mario de Brio, later confirmed that Silveira appeared to be in a daze and completely disorientated. Thinking that he had been attacked by robbers or bandits, Brio took Silveira to the hospital at Itaperuna and called the police. The police drove out to secure Silveira's abandoned car and look for clues.

The hospital found severe scratching and small cuts on Silveira's upper arms and discovered that his vision was blurred, but otherwise he was physically uninjured. The next day he was well enough to talk to the police. Only then did they realize that they were looking for a UFO and not a gang of violent thieves. They had found no signs of a fight by the abandoned car and soon gave up their investigations.

Silveira was never able to fill in what had happened to him while he was unconscious. His

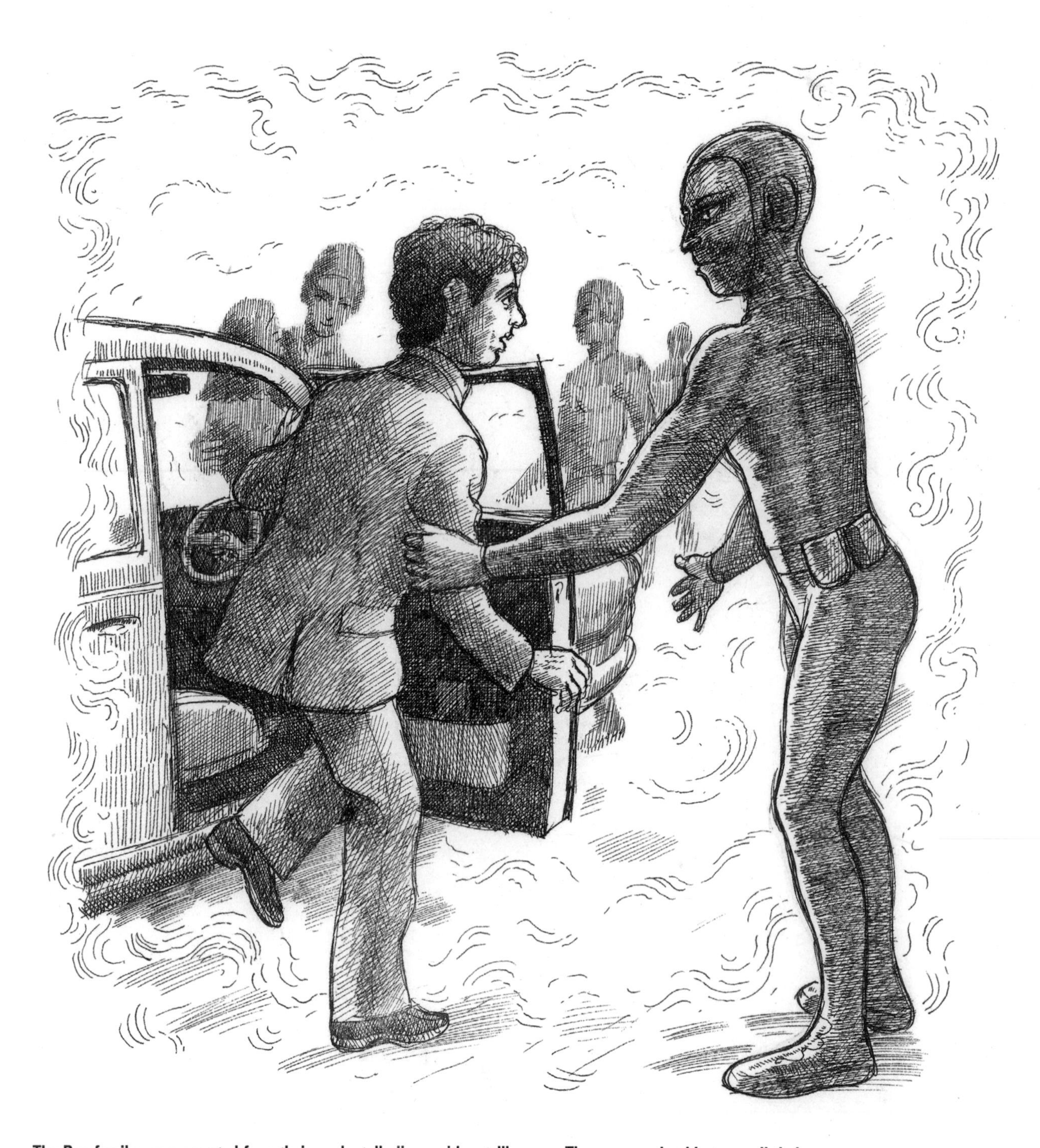

The Day family were escorted from their car by tall aliens with cat-like eyes. They were only able to recall their experience much later, under hypnosis.

sight returned to normal after a few days and the scratches healed completely. The only permanent damage was to his watch, which never kept good time again.

The Day family were also unable to remember much about their encounter in 1974. On 27 October the family were driving home to Aveley in Essex, England, when they sighted an oval blue light in the sky. The light seemed to be following them for a while, but then disappeared. Mrs Sue Day was keen to get home to watch a television programme at 9 pm, so they ignored the UFO and pushed on.

A few minutes later the car went round a bend and was engulfed by a green mist. The car radio at once began buzzing with static and John Day, fearing a short circuit, disconnected it. With a bump as if running over an object, the car came out of the mist and the Days drove home.

> **Wilder hypnotized both John and Sue to take them back in their memories to the night in question.**

While John stayed in the car to reconnect the radio, Sue took the children indoors and hustled them quickly up to bed. Then she went back downstairs to watch her TV show. To her astonishment the show was not on, the screen being filled by static. After fruitlessly trying to get a signal she glanced at a clock to see that it was almost midnight. It was at least 2½ hours later than it should have been. John could not explain the time lapse either.

In the weeks that followed all members of the family began to have nightmares in which there was a recurrent image of a monstrous face. The children were particularly terrified by the visions. Both John and Sue found themselves becoming increasingly interested in environmental issues, to which they had paid little attention before.

Eventually the nightmares and terrors got too much and the Days consulted their doctor. He sent them to see Dr Leonard Wilder who specialized in trauma and sleep problems. Wilder thought that something may have happened during the missing time on 27 October that had inflicted trauma of some kind and that the Days had subconsciously blanked it from their minds to help them cope.

Wilder hypnotized both John and Sue to take them back in their memories to the night in question. The Days independently recalled almost identical tales. When the car had entered the green mist the engine had first misfired and then died completely. Through the mist had then walked a group of aliens dressed in tight-fitting silver suits. These beings were over 1.8 m (6 ft) tall and generally human in appearance, except for the fact that they had cat-like eyes of penetrating power.

These aliens escorted the Days from their car through the mist to the blue UFO and led them on board. Neither John nor Sue could recall any form of physical force being used. It was as if they had wanted to go with the aliens perhaps in response to telepathic prompting of some kind.

Once inside the UFO the Days were handed over to a quite different group of aliens. These were about 1.2 m (4 ft) tall and dressed in white cloaks. They had animal-like faces with large eyes and pointed ears. It was these faces that had been haunting the dreams of the family. These creatures, the Days somehow realized, were inferior to the first group of aliens and had been trained by them to perform various duties. One of these was to conduct physical examinations of humans. The Days were then examined each in turn by the animal aliens.

The Days then recalled being led back out of the blue UFO and through the green mist to their car.

Once the medical tests were over, the taller aliens reappeared. They showed the family around the UFO, which was arranged on three decks. The aliens then told the Days that their home world was under threat of total devastation by runaway pollution and other environmental threats of a less specific nature. They showed them a holographic film of their home planet and the threats it faced.

Although it was not specifically stated, the Days got the impression that the aliens had been visiting Earth to study humans for some considerable time. It also seemed that the aliens were carrying out tests or experiments that involved genetics or children in some way.

The Days then recalled being led back out of the blue UFO and through the green mist to their car. John started the engine and they set off. The moment they came out of the mist their memories of the incident were wiped clean.

The alien abduction of police officer Alan Godfrey from outside the town of Todmorden in Yorkshire, England, on 28 November 1980 was in some ways similar to the Day case. He was in a rural area looking for some cattle that had been reported loose when he saw a UFO with a domed upperside and flat underside hovering over the road in front of his police car. He stopped his car and pulled out his notebook to draw a sketch of the object. Next instant he was parked by the side of the road about half a mile away with no idea of how he had got there, though a quick glance at the dashboard clock showed that it was now 20 minutes later.

Godfrey drove back to where he had seen the object to find that the road surface where the UFO had hovered was dry, although all around the ground was wet from recent rain. He later discovered that the sole of his boot was split.

After reporting the incident, Godfrey at first refused to talk to UFO researchers, but six months later agreed to hypnotic regression as he was by then keen to know what had happened during the missing minutes.

Under hypnosis, Godfrey told how a beam of light had shot out from the UFO as he prepared to sketch it. He had then fallen unconscious, coming to in a room lit dimly by some light source that he could not see. He was lying on his back on a hard table or platform. Beside him stood a figure that looked like an entirely normal human male aged about 50 and with a long beard. The man spoke to Godfrey in English explaining that his name was Yoseph, or something similar, and that he was not going to harm him, just carry out a few tests.

Godfrey then became aware of several smaller

A flying saucer swoops through a clear night sky in this artist's impression of a classic UFO sighting much like that of police officer Alan Godfrey in 1980.

figures that he took to be robots. These figures milled around him shining lights and prodding him with instruments that gave him severe head pains. When the robots backed off, Yoseph returned. He said that Godfrey could now leave, but that they would meet again at some point. Godfrey was then returned to his car, though he was not clear exactly how, and he then woke up as he remembered.

What these early abduction cases all have in common is interesting enough, though with hindsight the differences are just as arresting. First is the fact that almost all witnesses realized, after what appears to be a fairly standard UFO encounter, that there is a period of time which seems to be missing from their memory. They often then experienced nightmares or visions and some form of psychological upset. This led them to hypnotic regression in an effort to unlock the secret of the missing time.

This in turn led to the recovered memory of being abducted, more or less unwillingly, by aliens. Once inside the UFO the abductee was then usually, though not always, laid out on a table and had his clothes removed. The witness was then subjected to some form of inspection which he interpreted or was told was medical in nature. This inspection was sometimes, but not always excruciatingly painful. The witness was then spoken to by an alien which made protestations of friendship and peace. Finally the witness was taken out of the UFO and returned to whence they had come. Often the witness was rather hazy about how he got into or out of the UFO itself.

If this sequence of events set a pattern, it was broken by other elements. Some witnesses reported speaking to the aliens, others that conversation took place by telepathy. Crucially the aliens themselves seem to have been a very mixed bunch. Some were robots, others animal-like humanoids and some seemed indistinguishable from humans. Nor was there any real similarity between the inspections that took place nor the equipment used by the aliens.

UFO investigators were as puzzled by the differences of reported encounters as they were bemused by the apparent pointlessness of the tests the aliens seemed to be doing.

As the 1980s progressed, however, both these puzzles would fade. Both the aliens and the tests they carried out would begin to harmonize into a set pattern that began to be repeated again and again by witnesses.

One of the first reported abductions to fit what was to become a standard pattern was that suffered by 21-year-old Ros Reynolds and her boyfriend Corby outside Haverhill in Suffolk, England. The pair came across a UFO when driving one evening to visit friends. The UFO was hovering over electrical power lines and seemed to be sending out or receiving bolts of energy. Unnerved, Reynolds accelerated on her way.

Some witnesses reported speaking to the aliens, others that conversation took place by telepathy.

A few minutes later the car was bathed in an eerie blue light and the car engine and lights cut out. A serious silence ensued. Suddenly the car lights came back on and Reynolds restarted the engine. They drove on to their friends' house but found them asleep in bed. It was 1 am, not 9 pm as Reynolds and Corby had thought.

Although Corby suffered no ill effects, other than to be deeply puzzled by the events, Reynolds suffered a series of panic attacks that stopped her going out and she had recurrent nightmares that usually featured an evil face with large eyes.

After some months, Reynolds was sent for hypnotic regression therapy. This unlocked memories of a bizarre series of events that would soon become familiar to UFO researchers. She recalled the car being halted, then she was transported in some way she did not fully remember to a dim room which she somehow knew to be inside the UFO. There she was laid out on a table and studied by aliens.

These aliens were small with spindly arms, legs and bodies but with grotesquely large heads. The faces were dominated by large, jet black almond-shaped eyes that slanted upward to the sides of the head. The noses and mouths were small and the ears invisible. Researchers recognized this as a typical description of the alien type known as Greys. There was a larger, more human figure in the room, but it seemed to keep in the background. It seemed to be in charge, but this was not entirely clear.

The Greys swarmed around Reynolds as she lay prostrate and paralysed on a table. They were equipped with various types of probes and equipment with which they conducted an invasive

Several abductees report having recurrent nightmares featuring evil faces with large, black eyes that seem filled with menace.

gynaecological examination of a most unpleasant and unwelcome nature. When these prolonged studies were finally over, Reynolds was returned to her car, again by means that she could not clearly recollect.

The hypnotic therapy did help Reynolds, but it was months before she was once again able to lead a normal life as she had done before her nightmarish encounter.

Fitting into a similar pattern was the abduction of Julio Platner from a rural road near Winifreda, Argentina, on 9 August 1983. It was not yet dark when he saw a bright, bluish light hovering over the road ahead of him. Platner stopped his van, and got out to try to get a better look at whatever the object was. He was suddenly hit by a beam of brilliant light that dazzled him and sent him stumbling backwards. He remembered falling, then came to lying naked on a flat, hard table in a dimly lit room. The room was about 3 m (10 ft) across with curved walls that glowed slightly.

Moving about around him were four humanoid creatures. The figures were basically human in shape, though their limbs and bodies were slender and rather elongated. They stood just over 1.2 m (4 ft) tall and were a pale greyish-white colour all over. Platner was not certain if they were naked or dressed in very tight clothing. Their heads were entirely bald with domed foreheads. Their noses, mouths and ears were tiny, but their eyes were

An artist's impression of a typical abduction. The aliens are 'Greys' and use medical instruments of some kind. The humans usually report being unable to resist the short aliens.

huge and bulging without any eyelashes.

In other words, Platner had been abducted by aliens fitting into the Grey category. For some reason that he could not really explain Platner thought that one of the Greys was a female.

> Platner then received a telepathic message telling him not to worry as the aliens would not harm him.

Platner tried to sit upright, but was held firmly in place by one of the Greys gripping his shoulders. He tried to scream, but no sound came from his mouth. One of the Greys came over to stare at Platner with its huge black eyes. Platner then received a telepathic message telling him not to worry as the aliens would not harm him. 'What you are experiencing now,' the Grey continued, 'has happened to thousands of other people. When it is over you can talk of this if you like. Some people will believe you, but most will not.'

After this communication, Platner became calmer and more relaxed. The female came over with a long, rigid tube which she filled with blood from Platner's arm. Further tests and medical examinations followed, lasting about 20 minutes. Platner was then told to stand up and was given his clothes and belongings. He got dressed, then passed out once again. When he recovered consciousness he was lying on top of the roof of his van a mile or so from where he had encountered the UFO. On his left arm was a small wound at the spot where the female Grey had extracted blood.

In 1988 Jim Weiner sustained a head injury and during the medical recovery process he underwent hypnotic procedures. This led in turn to the admission by Weiner that he, his brother and two friends had seen a UFO when on a camping trip as friends back in 1976. Jim's brother Jack, Chuck Rak and Charlie Foltz all subsequently agreed to be questioned about the experience.

They each recalled clearly that they had been canoeing and camping in the Allagash River area for some days in August 1976. On 26 August they had camped beside East Lake and, as dusk fell, paddled out to fish the lake waters to catch their supper. They saw a large, round object hovering some distance off to the southeast. When Foltz switched on an electric torch, the object had swooped towards them. A beam of light flashed out in the direction of the young men. The next thing they could remember was standing on the lake shore about 3 hours later. The UFO was flying off.

Under separate hypnotic regression each of the men told a broadly similar story of what had happened during those missing hours.

The men recalled that as the beam of brilliant light hit the canoe, they had panicked, then been overwhelmed by lethargy and tiredness. They had then floated upwards along the beam of light towards the UFO. Passing into the craft they had been confronted by a group of aliens matching the

Although descriptions of UFOs can vary a good deal, those involved with abductions tend to be round, domed on top and bottom and between 7.5 m (25 ft) and 23 m (75 ft) in diameter.

usual description of Greys. The Greys had used some form of mind control directed through their eyes to force the men to undress.

The men were then sprayed with a fine mist-like gas before being forced to lie down on tables. They were then subjected to a series of studies and

They were then subjected to a series of studies and investigations of an apparently medical nature.

investigations of an apparently medical nature. Painful probes were inserted into their bodies to extract samples of blood, skin, urine and sperm. Throughout all this the Greys seemed utterly indifferent to the feelings and pain experienced by the men.

The men were then allowed to get dressed. The Greys returned and, again using telepathic coercion, forced the men to walk through a circular doorway. They were then engulfed by the light beam once more and transported back to their canoe. The students paddled ashore, clambering out as the UFO took off, which was when their conscious memories took over.

By the late 1990s the alien abduction experience was being reported by hundreds of people. Although details varied, the various reports fitted into a general pattern.

Most abductions begin with the arrival of the aliens. Typically the witness first sees a UFO, though this is not always the case. Some abductees are taken from quiet, rural areas, others from their own homes. If a UFO is not seen, the witness normally reports that he has seen something very unusual, most often humanoid entities. This phase

A 'Grey' alien looms out of the mist. Several witnesses have reported that a fog or cloud envelops them as the abduction takes place.

Once on board the UFO, abductees are sometimes given a tour of the craft. Betty Hill (see Encounter Casebook No. 10 on page 182), was shown a star map, leading to much speculation about the home planet of the aliens who abducted her.

takes place in a perfectly normal and natural setting.

The witnesses are then abducted. This may involve a degree of physical force, but more often the witness reports being unable or unwilling to resist even though they feel deep unease or fear about what is happening. It is as if some telepathic coercion is taking place.

Once abducted the witness finds himself or herself in a room lit fairly dimly and often with no obvious light source. Very often the witness is fairly hazy about quite how they got there. The witness is either already on a table or slab, or is quickly forced on to one. The aliens – usually of the Grey variety – then close in to conduct assorted medical examinations causing varying degrees of pain and unease. The aliens are often said to be particularly interested in the reproductive system, taking sperm or egg samples. The abductee may then be scanned or surveyed by complex machinery.

When the examination is over, the abductee is often then approached by an alien leader. This authority figure may be a Grey, but is often either significantly taller than its fellows or much more human in appearance. This leader has a conversation with the abductee that usually involves apologies for the way they have been treated and a chat about the UFO and how it works. The abductee is sometimes taken on a tour of the craft.

Women are sometimes taken to a nursery or medical room containing infant human-alien hybrids. They are encouraged to hold or cuddle the odd beings. Many women who experience this get the strong feeling that the babies are sickly or ill and in desperate need of human warmth and love – the Greys are a notoriously cold and emotionless race.

The abductee is then returned. Most are usually put back in the same place from which they were abducted, but not always. At this point the conscious memory is often wiped clean. There is, however, usually some sign that something odd has happened. The witness might have minor cuts or scratches or have his shirt on backwards. There is usually an apparent period of missing time during which the witness cannot recall what has happened.

In the following days or weeks the witness will often suffer from nightmares, depression and other assorted sicknesses. This may lead the witness to consult a doctor or, less often, a UFO research group. Hypnotic regression may then unlock the memory to reveal what has happened.

It is likely that many people who experience a period of missing time dismiss the event from their minds, especially if they subsequently suffer no adverse reactions. There may, therefore, be many more cases of the alien abduction experience than is known to researchers.

Seemingly divorced from this growing trend towards uniformity in alien abduction experiences was that of Alfred Burtoo on 12 August 1983.

Burtoo was sitting enjoying a quiet evening's fishing beside a canal in Berkshire, England, when he saw a light circling overhead. Assuming this to be a helicopter as he was close to an army base, Burtoo took little notice. He poured out a cup of tea from his flask as he watched the 'helicopter' land a short distance away.

In the following days or weeks the witness will often suffer from nightmares, depression and other assorted sicknesses.

Two short men left the object and began walking towards Burtoo, whose dog at this point got up and began snarling. The figures were about 1.5 m (5 ft) tall and dressed in green overalls with headgear rather like motorcycle helmets which had visors covering the face. The two men asked Burtoo to accompany them. Burtoo put down his tea and followed them.

He climbed a short flight of steps into the UFO, having to duck his head as the doorway was very low. Once inside the craft, Burtoo was bathed in an orange light for a few seconds. Then one of the figures asked him how old he was. Burtoo replied that he was 77, which was the truth. There was a hurried conversation between the two ufonauts after which one of them announced: 'You can go

now. You are too old and infirm for our purposes.'

Burtoo was hurriedly pushed out of the craft. The door was slammed shut behind him and the UFO began to rise silently into the air as Burtoo went back to finish his cup of tea, which was now rather lukewarm.

Burtoo's experience was certainly atypical. The aliens seemed more like Tricksters than Greys and he was not subjected to any painful or invasive medical procedures. More strikingly he could recall all aspects of his encounter with no need for hypnotic regression.

> ... strikingly he could recall all aspects of his encounter with no need for hypnotic regression.

Some researchers, however, have interpreted Burtoo's encounter as actually reinforcing the general trend. They point to the aliens' insistence that he was too old for their purposes. If the unstated purposes were related to the human reproductive system then a man of 77, albeit a fairly active one, might be deemed too old. The fact that he was rejected might mean that he was thrown off the UFO before he could suffer the unpleasant experiments or meet the Greys that might have lurked further inside the UFO.

In 1983 Alfred Burtoo was the subject of what seems to have been an attempted abduction, though the aliens lost interest when they realized how old Burtoo was.

Encounter Casebook No. 10

TYPE>>**Alien abduction** DATE>>**19 September 1961** PLACE>>**Highway 3, near Lancaster, New Hampshire, USA**
WITNESSES>>**Betty Hill, Barney Hill**

Although their encounter took place in 1961, the Hills did not hit the public headlines until the publication in 1966 of a book based on their experiences. *The Interrupted Journey* was an immediate bestseller and its contents revolutionized research into UFOs and aliens.

The Hill encounter began when Betty and Barney were returning from a holiday in Canada. They were driving down Highway 3 hoping to arrive home in Portsmouth, New Hampshire, around 3 am. At about 11 pm, when passing Lancaster, the couple saw a bright light in the sky that seemed to be flying north at high speed. It then turned suddenly and moved southwest at a slower pace. The Hills drove on for a few miles, with the object seeming to keep pace with them, though it was moving in a most erratic manner.

About halfway between Lancaster and Concord, the Hills pulled over and got out of their car to watch the object. It came down closer. They could now see that it was an oval-shaped bluish-white object with a row of windows on its leading edge and two projections on either side. Through the windows could be seen diminutive humanoids trotting about as if attending to some mission or other. A red light appeared on each projection.

Barney Hill suddenly took fright. He insisted that his wife get back in the car as he was convinced that 'they' were going to attack. The Hills got back in their car and drove off at high speed. Although they heard beeping noises and what seemed to be stones hitting their car, they soon left the UFO behind and drove on through the night. They got home around 5 am and went to bed.

Five days later Betty Hill wrote a letter to UFO investigator Donald Keyhoe outlining what had happened. A few weeks later Keyhoe and his team contacted the Hills and on 25 November visited to get a detailed version of the encounter. It was during this process that the Hills realized that there was a time discrepancy. They had stopped for only a few minutes to look at the UFO, and yet were 2 hours late getting home.

The investigation was detailed and careful. It concluded as follows: 'It is my view that the observer's blackout is not of any great significance. I think the whole experience was so improbable and fantastic to witness that his mind finally refused to believe what his eyes were perceiving and a mental block resulted.'

The matter did not end there. Barney's account was soon found to be riddled with inconsistencies. Although he was adamant that he had seen the

The Hill abduction began as a UFO followed the Hill automobile along a remote country road in New Hampshire. The couple were able to give a clear description of the craft, but lost their memories of what happened next.

crew of the UFO only at a distance and indistinctly, he also reported that the leader of the crew had an expressionless face and that another crew member had grinned at him when looking back over his shoulder.

It was Betty, however, who was to have real problems. Over the following months she had a series of vivid nightmares. These contained disjointed and terrifying images linked to the UFO encounter. She saw herself undergoing very painful medical tests, being marched into the UFO by uniformed aliens, talking in English to the aliens, and being shown a star map. The jumble of scenes emerged gradually over a long time.

Barney meanwhile began to suffer from depression and lethargy as well as a nasty stomach ulcer. His doctor thought that he might be suffering from mental issues related to accumulated past experiences that had become too much for him to cope with. Barney was therefore sent to see Dr Benjamin Simon who specialized in hypnotic regression to expose and therefore deal with the causes of emotional problems.

After a few sessions with Barney, Dr Simon asked Betty to attend a session. After several hypnotic regression appointments with the two separately, Dr Simon revealed what he had found. The results seemed to explain both Betty's nightmares and Barney's problems. Under hypnosis, the Hills had told almost identical stories.

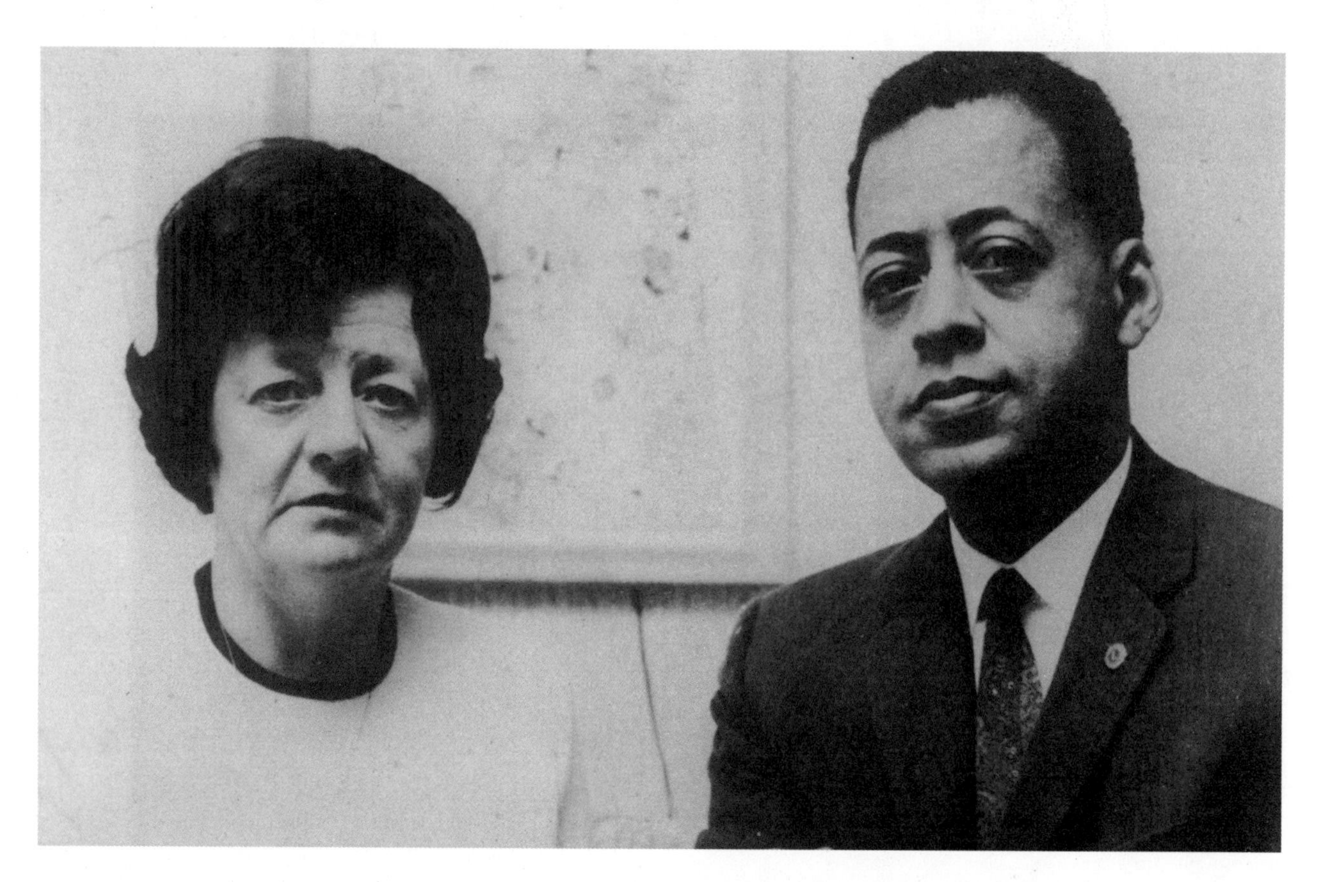

Betty and Barney Hill photographed at the time of their abduction. The aliens seemed fascinated by the difference in skin colour between the couple.

According to the memories unlocked by hypnosis, the Hills had seen the UFO much as they recalled. But as Barney drove off, the car was stopped by a dozen men standing in the road. The men had opened the car doors, reached in and pulled the Hills out. The Hills, who were strangely unable to struggle or resist although both were very frightened, were then escorted by the men into the woods and to a clearing where the UFO was resting on the ground.

The ufonauts were described as being around 1.2 m (4 ft) tall and dressed in tight-fitting suits of black leather or some other smooth fabric. Some of them had small peaked caps on their heads. The humanoids were basically human in proportion, but their heads were large and oval-shaped with pointed chins, snub noses and slit-like mouths. There did not seem to be any external ears. Their eyes were large and slanted, but otherwise fairly human in appearance.

The Hills were then marched up a ramp into the UFO itself. Barney's recollections became vague at this point, but Betty was able to recall events clearly. She was taken to one room and Barney to another. In the room Betty met an alien who could speak good English. He asked Betty questions about her age, diet and physical health and then took samples of ear wax, hair, fingernails and skin scrapings.

The man then asked Betty to lie on a couch and attached her to a large machine equipped with needles, dials and gauges. He removed Betty's dress and began an extremely painful examination that involved jabbing needles into Betty, apparently to see if she was pregnant. Another of the aliens then entered and waved his hand, at which the pain left Betty. The new arrival then asked to look into Betty's mouth and seemed surprised that she could

Betty explained that Barney had false teeth. This led to a conversation about the ageing process in humans.

not take her teeth out. Betty explained that Barney had false teeth. This led to a conversation about the ageing process in humans.

The new arrival, whom Betty took to be the leader of the crew, then talked to her for some time, seemingly fascinated by the fact that Barney had black skin and Betty white. He also showed Betty a map of the location of the star system where he said they had come from. Betty asked for some proof of the aliens' existence and the leader handed over a book, but it was soon taken away when other members of the crew objected. The leader then told Betty that neither she nor Barney would remember anything of the event adding that even if they do nobody will believe them.

The Hills were then marched back down the ramp and returned to their car. The UFO took off and the Hills continued their journey home. The first alien abduction to become public was over.

CHAPTER 8

What is Out There?

The truth, so the saying has it, is out there. But what is the truth and, more specifically what are people reporting when they witness a UFO or alien encounter? Therein lies one of the central problems of the entire UFO and alien encounter enigma. We are, by and large, dealing with what people report that they have seen and what has happened to them. Nobody has ever publicly produced an alien being or an alien spacecraft. There are occasional marks left by UFOs and aliens in passing, but no direct and incontrovertible evidence.

It is also important to recognize that the UFO experience is not entirely new. It was Kenneth Arnold's sighting in 1947 and the widespread publicity it gained that pushed UFOs, or flying saucers as they were then known, into the public arena. But reports of UFOs and ufonauts have been around for centuries. These pre-1947 reports were generally interpreted as sightings of fairies, goblins, trolls or angels. Those inclined to believe in such entities took the reports as fact, those who did not dismissed them. Clearly whatever the truth may be, it has been around for a very long time.

An actor dressed and made up to play the role of a goblin. The similarities between the reported appearance of aliens and of goblins and other folkore entities can be striking.

Probably the most productive first step is to be clear over what we are talking about. As we have seen in the course of this book, there are several interrelated phenomena to be covered.

Some of these objects are seen at remarkably close quarters and a few have a direct and observable impact on their surroundings.

In the first instance there are flying objects that seem to be neither natural nor man-made. There is no acknowledged, rational explanation for these flying objects, which generally go by the name of unidentified flying object, or UFO. These objects are sometimes seen during the day, sometimes at night. The objects may be picked up on radar, but often are not. Some of these objects are seen at remarkably close quarters and a few have a direct and observable impact on their surroundings.

Seen in connection with some of these objects are a variety of apparently intelligent life forms. Most of these are humanoid in form – having two arms, two legs and one head – but a few are not. Sometimes these entities ignore the humans who see them, sometimes they react. On a few occasions the creatures are reported to communicate with the humans who encounter them. In the alien abduction experience the communications are not always pleasant and the results for the humans can be distressing in the extreme.

There is nothing to prove that all these phenomena are related to each other. The strange lights seen in the sky at night are not necessarily the same things as the discs and other UFOs seen by daylight. Nor is it certain that all the flying objects contain life forms. Likewise, we cannot be sure that the entities arriving in UFOs are those that carry out abductions.

The various elements to the enigma have become linked largely because they seem to be so. But when evaluating the evidence it is as well to keep in mind that we might well be dealing with two, three or even more quite distinct phenomena.

It must be accepted that many sightings of UFOs are of perfectly normal things which are not recognized for what they are.

A clear example of this occurred over central England in 1990. Several dozen witnesses saw large, dark objects floating overhead at low speed accompanied by bright flashing lights and occasional shafts of an intense light beam. These were reported as UFOs and made the local press. So frequent did these sightings become and so reputable were the witnesses that the local Member of Parliament felt moved to ask a formal question in the House of Commons to see if the authorities could investigate.

The truth of what was happening was kept secret at the time, but came out later. The dark shapes, flashing lights and searchlight beams had been caused by formations of bombers practising mid-air

Seen from the cockpit of a fighter jet, the boom of a refuelling tanker comes down to make contact in mid-air. Similar manoeuvres at night led to a rash of UFO reports from England in 1990.

refuelling at night. This manoeuvre is tricky enough by daylight, but extremely difficult when it is dark. The reason for the secrecy was that the bombers would be used to attack Iraq during the liberation of Kuwait and the RAF and USAF wanted to keep secret the fact that they were perfecting this technique to avoid alerting the Iraqis to the fact before the air strikes were to be launched.

Tucked away in the deserts of Nevada is a US military base known variously as Area 51 and Base S4. Only those with the very highest security clearance are allowed to enter the site. The US military has bought a vast area of land around the base, including all hills that overlook it, and the entire area is closed off to the public. Armed guards roam around and the security fencing is impressive. It is here that the USAF develops and tests its latest technology. There can be little doubt that some of the UFOs seen performing apparently impossible stunts in the area are, in fact, secret aircraft equipped with ultra high-tech features.

Rather more dramatic, but completely natural, was the ball of fire that raced through the skies over the Komovi Forest in Yugoslavia on the evening of 26 November 1967. The object was seen by a group of forestry workers who were sheltering from a sudden downpour of heavy rain. The object streaked through the sky, then plunged into the forest and set the trees alight. The resulting forest fire was put out with some difficulty despite the wet weather. Almost certainly this was an example of ball-lightning – a rare electromagnetic phenomenon which occurs when electrically charged air forms a white-hot ball during a thunderstorm.

> It must be accepted that many sightings of UFOs are of perfectly normal things which are not recognized for what they are.

A number of USAF jets parked on the concrete of Nellis Air Force Base in Nevada. Some suspect that the testing of top secret military aircraft can explain some of the UFO reports made each year.

Other apparently genuine UFOs turn out to be deliberate hoaxes. In 1962 a 14-year-old British schoolboy, Alex Birch, photographed five saucer-shaped objects flying over his home in Sheffield. He sent the photo to the local press, who passed the story on to the air ministry and to national media. Soon Alex was being interviewed on radio, television and by the authorities. He stuck to his story and convinced nearly everyone who spoke to him. Photographic experts found no evidence of trickery or manipulation in the photographic negative.

He had produced the photo by painting the 'discs' on to his bedroom window and then taking a photo through the window of the scenery beyond.

More than 10 years later, Alex admitted that the whole affair had been a hoax. He had produced the photo by painting the 'discs' on to his bedroom window and then taking a photo through the window of the scenery beyond. What were really small paintings close up seemed to be large objects far away. Alex claimed to have taken the photo as a joke, but to have become frightened when government officials arrived and so was too scared to admit that the photo was a fake.

It is sometimes possible to recognize frauds, advanced technology and unusual natural phenomena for what they are, but not always. There are probably hundreds of cases of seemingly baffling UFOs which are perfectly normal things, but for one reason or another remain unidentified.

There is still, however, a core of reports that cannot be so easily dismissed. These are objects seen by reputable and reliable people in clear conditions at short range and which very often have a physical impact on their surroundings. Each one of these on its own might be dismissed, but together they form an impressive body of evidence.

It is worth looking at these reports to see what they tell us about the phenomenon, or phenomena, being reported.

Firstly, the objects usually fall into a size range of between 6 m (20 ft) and 18 m (60 ft). Some that are smaller are reported and some larger, but only rarely.

Secondly, while their speed in flight might be fast or slow, it is often reported to be rather erratic. Witnesses say that UFOs wobble or rock from side to side, follow winding paths and zigzag about. They are also likely to hover, again while wobbling or rocking slightly. UFOs may also appear apparently from nowhere, and then vanish as abruptly as they arrive.

Thirdly, the objects will very often emit a sound that has been likened to a hum, whine, whistle or whoosh.

When seen in daylight the objects are usually described as being silver in colour and metallic in composition. Witnesses report that they are polished, shiny or glittering. When seen at night, the objects tend to glow across a spectrum of colours from white through blue, green and orange to red.

A photo taken in 1958 that appears to show UFOs flying around the Eiffel Tower in Paris. In fact the discs are the reflections of café lights as the photo was taken through a plate glass window. Not all cases of photographic effects are so easy to spot.

Finally the flying objects may have a physical impact on things around them. Electrical systems are prone to fail when a UFO is nearby – radios and car engines often cut out quite suddenly. Witnesses often report feeling a radiant heat as a UFO passes and signs of high temperatures are sometimes shown by vegetation and soil around where a UFO has come to ground. In a few instances it seems that something akin to radiation is expelled by the UFO.

What all this means is open to question, but the key thing to recognize is that a vast number of sightings and encounters repeat the same features time and again. The conclusion seems inescapable that what witnesses are reporting is a real phenomenon. What that phenomenon is, however, is another matter.

Those who seek a natural solution to the mystery theorize that these UFOs may be a freak and very rare electromagnetic event.

Those who seek a natural solution to the mystery theorize that these UFOs may be a freak and very rare electromagnetic event. Certainly the humming sound, impact on electrical equipment and other elements do indicate that this may be the case, or at least that strong electromagnetic fields are somehow involved. However, no known electromagnetic event could produce such effects and until one is found the idea remains merely a hypothesis.

Moreover, the idea that these UFOs are natural atmospheric events does nothing to explain the intelligent entities that are so often seen inside them or standing close by. These beings are seen by reputable and reliable witnesses, so there is no reason to believe anything other than that the witnesses genuinely believe that they have seen what they report.

What these reports suggest is that the UFOs are being used as transport by intelligent beings that are not human. These beings come in a wide variety of shapes and sizes. Noticeably the vast majority of them are humanoid. They may differ from humans in size and detail, but they are all basically built to the same body plan. There are some that break this mould – looking like disembodied brains, bowls of jelly or monstrous birds – but they are very much in the minority.

The behaviour of these ufonauts varies as widely as does their appearance. They may be intent on collecting things such as water, plants or animals, or they may appear indifferent to their surroundings. They may flee when they realize that a human has seen them, or they may attack. Some can talk human languages, others can communicate telepathically. But other entities cannot make themselves understood; they simply babble on in a strange tongue or wave their arms about as if using a form of sign language.

When they can make themselves understood the ufonauts impart a variety of messages ranging from the obscure to the nonsensical. Those that choose to describe their craft and their mission invariably deal in platitudes or speak with such vagueness that their comments are meaningless. None of them has ever handed over a blueprint for how their craft operates, instead they babble on about condensed electricity, gravity condensers and the like. Sometimes humans who encounter these entities are treated well, but others are subjected to bizarre ordeals.

Quite clearly the ufonauts do not all have the same origin or explanation. Dog-faced midgets cannot be confused with 1.8 m (6 ft)-tall blond athletic types. While UFO researchers can fit most sightings of ufonauts into one of a small number of categories based on their appearance and behaviour, that does not get us any closer to understanding who or what these entities might be.

When looking for explanations of the UFO and alien encounter enigma, most researchers have settled on variations of a small number of themes.

Those who deny the reality of the entire experience argue that none of the reported encounters can be taken at face value. Some witnesses, they say, are just lying. Others are revealed to have psychological conditions and may have imagined the entire event. Even those without any psychological issues are capable of generating hallucinations in the right conditions, and those hallucinations can be terrifyingly real to the person having them.

These sceptics point to several facts about the reports that support their ideas. For instance there is the undeniable fact that the type of aliens being reported has varied over time. During the 1950s and 1960s many witnesses reported seeing the type of ufonaut classified as Nordic. These beings are basically human in almost all respects – except being impossibly good-looking. They are well-intentioned towards humanity and often issue warnings about imminent catastrophe. More recently the majority of witnesses report encountering the hostile Greys.

The sceptics then point out that blockbuster movies of the 1950s tended to depict aliens as being basically human – usually due to problems with special effects. Since the 1970s, however, films have been able to be more adventurous in their depictions of aliens (see picture overleaf) and have been especially fond of creatures that look like the Greys. This much is true, but it ignores the fact that the movies picked up the image of the Greys from earlier witness reports. In any case the idea that all UFO and alien encounters are hallucinations is not borne out by physical evidence, nor does it take account of the fact that people situated thousands of miles apart who have never met each other nor read about each other's sightings report basically similar events.

Jung postulated that the human brain was hardwired with what he called the collective unconscious.

One theory that deals with the main problems of the hallucination explanation, though it has problems of its own, is the psychosocial hypothesis. This has its roots in the work of the Swiss psychologist Carl Jung. Jung postulated that the human brain was hardwired with what he called the collective unconscious. This meant that humans are likely to react in similar ways to similar events, drawing on underlying processes inherited from our ancestors. This idea holds that people interpret events according to the prevalent ideas of their time and culture.

A scene from the 1977 movie *Close Encounters of the Third Kind.* The film drew on many genuine reports of UFOs and ufonauts to create an entirely fictional storyline that nonetheless reflected the reality of UFO reports.

One case that demonstrates how much of the alien encounter phenomenon is in the mind of the witness occurred at the Swedish village of Vallentuna in 1974. On 23 March a young man named Anders was walking home from work in the dark. As he strolled along a couple of miles from the village, a bright beam of light shot out from the darkness and seemed to blast him backwards, lifting him from his feet. Anders lost consciousness.

The next thing that Anders remembered was being cradled in his wife's arms just outside the doorstep of his house. His body was pocked by severe burns, but he could recall nothing of what had happened. His wife had heard a thump against the door and opened it to find Anders slumped on the doorstep.

The incident was reported to the police, who treated it as assault, and Anders was taken to hospital. The police investigation failed to turn up any real clues, but they did find a woman who had seen the beam of light, but had not seen Anders. Once his burns had healed, Anders agreed to undergo hypnotic regression to see if he could recover any memories of who had assaulted him and why. The exercise was undertaken by Dr Ture Arvidsson, who was experienced in helping the police and in using the technique to allow patients with emotional problems to confront the cause of their difficulties.

Under hypnosis, Anders recalled that after being struck by the beam of light he had floated through the air as if levitating. He had then passed through a doorway to enter a dark room in which was a group of human-like creatures. These creatures then used probes and other instruments to take samples of Anders' skin and apparently subject it to various tests.

The beings then told Anders that they were going to wipe his memory so that he would recall nothing about them. First, though, they asked if he would like to be dropped off at home. He said he would, and was sent back out of the room on the beam of light to arrive at his front door.

What is interesting about this case is that Anders at once assumed that he had been kidnapped and tortured by trolls. These trolls are prominent figures in Scandinavian folklore. They are supernatural beings that are said to live beneath the ground, able to come out only when they run no risk of being hit by direct sunlight which would turn them to stone. They are generally indifferent to humans, but if offended they may turn belligerent and hostile. Kidnapping and torturing humans would seem to be a favourite occupation for trolls who feel themselves offended by people for some reason. His doctor, on the other hand, thought that he was suffering from deep rooted problems and that his subconscious had invented an encounter with trolls to rationalize those hidden issues.

It was only when the story was reported outside his local area – where it was treated as a troll encounter – that UFO researchers noticed its basic similarity to alien abduction incidents. Although Anders had seen no UFO, everything else fitted into the general pattern.

A garden figure of a troll. These Scandinavian folklore creatures share many similarities with modern ufonauts. In 1974 a Swede reported a meeting with trolls that showed all the marks of an alien abduction.

In 1869 an Englishman had a terrifying encounter with boggles, a form of malevolent fairy, which fled when he brandished a Bible at them. The boggles behaved much like modern 'Grey' aliens.

Much the same variation in interpretation of an event could be the case with a Mr Brown from Bewcastle in England's Lake District. Brown was riding from Lanercost to his home one evening in 1869 when he saw a group of short humanoids in the road in front of him. The figures were described as being around 0.9 m (3 ft) tall and dressed in dark clothing. They had large heads with pointed chins and excessively large, pointed ears.

As soon as the humanoids saw Brown, they attacked him. They scratched his face and clothes with their claw-like hands and dragged him off his horse. The figures began dragging Brown towards a glowing mound beside the road. After a good deal of struggling, Brown managed to pull an old Bible out of his pocket and brandish it at his assailants. At this, the humanoids gave up the attack and fled into the mound through a door that opened in its side.

Brown fled home. He ascribed his adventure to an encounter with boggles. Like the Swedish trolls, these boggles were an established part of local folklore. They were believed to be a type of fairy that was at best mischievous, and at times malevolent towards humans. They delighted in playing tricks, some of them very dangerous to the humans involved.

Reading the account of Brown's encounter now, his adventure seems more like those of the Sutton family involved in the Kelly-Hopkinsville alien battle than anything else.

Even if Jung's ideas are accepted, they do not fully explain what it is that people interpret in these various ways.

He ascribed his adventure to an encounter with boggles. Like the Swedish trolls, these boggles were an established part of local folklore.

Far and away the most popular theory is the extraterrestrial hypothesis, or ETH. Put simply this holds that the entities seen with UFOs are alien beings from another planet and that the UFOs are their means of transport. The idea took hold almost immediately after the 1947 Arnold sighting, most prominently in the books by Donald Keyhoe. It has remained the most popular explanation ever since.

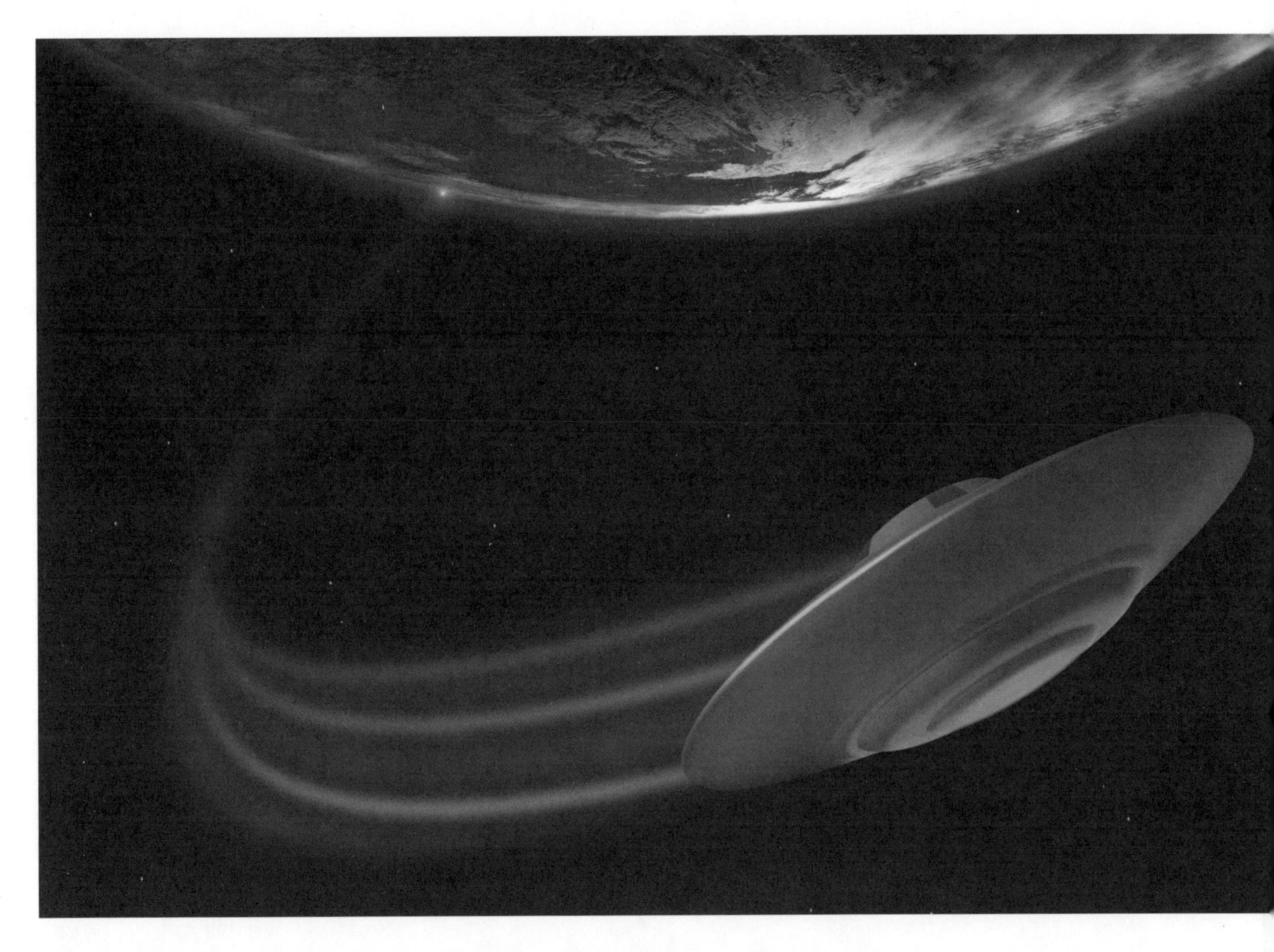

An artist's impression of a UFO bursting out of Earth's atmosphere. The idea that UFOs are alien spacecraft is perhaps the most popular explanation of the UFO phenomenon among the general public.

Certainly it would explain almost everything about the UFO and alien encounter experience if the ETH were true. Our own technology is so superior to that of five centuries ago that it is easy to believe that another civilization more advanced than our own could produce craft that perform the way UFOs do. And the reported ufonauts are quite obviously unlike anything on Earth.

There are, however, problems with the ETH. One of the most pressing would be the sheer distances involved in interstellar travel. The vastness of space is almost unimaginable. Sending a spacecraft around our own solar system would take years, to send one to the nearest star to our sun would take decades. To get to the nearest known sun with planets would take centuries. Even assuming that our current rocket technology is fairly primitive, there seems to be an absolute maximum speed for travel that is around the speed of light. We have no way of reaching such speeds, but even if we did, interstellar travel would involve spending years sitting on a spacecraft doing nothing at all before

> We have no way of reaching such speeds, but even if we did, interstellar travel would involve spending years sitting on a spacecraft doing nothing at all...

arriving. It seems scarcely credible that aliens would go through all that effort simply to land on Earth, collect a few plants, conduct some medical tests on humans and then fly off again.

However, the fact that modern scientists tell us that it is impossible to travel faster than light does not mean that this is so. In the past scientists have stated with absolute certainty, backed up by reams of data and proofs, things that we now know to be untrue.

The Mars Polar Lander, a space probe sent to Mars by NASA to investigate the planet. Some theorize that ufonauts are on a similar mission to Earth.

For many years it was a well-known 'fact' that no life could exist in the oceans below a depth of about 600 m (2,000 ft). Then, in 1860, an early submarine telegraph cable was hauled up from around 3,000 m (10,000 ft). It was found to be encrusted with clams, corals and other living things. The reaction of the marine scientists was instructive. They discounted the evidence on the grounds that the men who had hauled it up were not scientists. It was not until 1872, by which time the evidence was overwhelming, that scientists accepted the truth.

Most readers will know that in the early 21st century climatologists are predicting global warming, rising sea levels and assorted disasters due to mankind's pollution of the atmosphere. Fewer will know that in the 1970s climatologists were predicting an imminent ice age and sea level drop for precisely the same reasons.

The problem is not that scientists are wrong, but that they are prone to describe as a fact something which is really a theory that nobody has yet managed to disprove. Perhaps travel faster than the speed of light is impossible. Or perhaps we just have not yet worked out how to do it.

Of course, many ufonauts in their conversations with humans claim to have come from other planets. These origins are often referred to only vaguely with no indication of where they are. At other times the aliens claim to come from Venus, Mars or another planet in our solar system. Space probes have since shown these planets to be devoid of life, so such claims are clearly false.

One exception comes from the Hill case (see Encounter Casebook No. 10 on page 182). Betty Hill was shown a star map that she subsequently redrew under hypnosis. The map appears to show the aliens' point of origin to have been the Zeta Reticuli star system. This system has been studied since the Hill encounter and is now known to contain two stars very similar to our own sun, raising the possibility of life on planets that orbit them. Quite why aliens able to travel interstellar distances needed a paper map is not so clear.

> **The ultraterrestrial hypothesis ... holds that the UFOs and their occupants come not from another planet but from another dimension.**

Another idea goes by the name of the ultraterrestrial hypothesis, or UTH. This holds that the UFOs and their occupants come not from another planet but from another dimension. The existence of such dimensions does have a theoretical basis in modern physics, but they are little understood and there is no known way that they could be linked or that travel between them could take place. Which is not to say that such interdimensional travel is impossible.

The time travel hypothesis, or TTH, speaks for itself. This holds that UFOs are not devices for travelling in space, but for travelling in time. They

A tribal shaman. Shamans typically use self-induced trances to visit the spirit world and converse with gods and demons. Some believe that encounters with aliens are similar experiences.

are thought to be travelling back to our own time from some remote future. They may come to conduct historical research, to harvest genetic material from plants or animals extinct in their time or for some other purpose.

Rather less popular in recent years has been the hollow earth hypothesis. This holds that the UFOs and their crews come from an advanced civilization that exists deep beneath the surface of Earth. The idea seems rooted in Brahmin legends current in India, supported by other folklore and legends. There seems to be little real evidence to support this idea, other than the fact that it gets around many of the problems of the ETH.

The shamanic hypothesis takes its name from the shamans that exist in many pre-literate societies. These people are those who are able to communicate with the spirit world that is all around us, and yet is usually invisible. This idea says that there is an underlying consciousness of the universe that breaks through to human minds by way of shamanistic rituals leading to encounters with spirits or, given our increasingly technological society, by way of supposed encounters with aliens. It is probably fair to say that this idea has few adherents.

Another minority view, though it has been gaining some ground since it was postulated by the French UFO researcher Pierre Guérin, is the induced dream hypothesis, or IDH. This idea takes as its starting point that many encounters with aliens have similarities that would seem to indicate that there is a reality to the experience and that they are not mere hallucinations. It also takes into account the absurdity of many reports, the apparent pointlessness of alien activity and the fact that many sightings of UFOs are unsupported by other people nearby.

> The hollow earth hypothesis ... holds that the UFOs and their crews come from an advanced civilization that exists deep beneath the surface of Earth.

Guérin suggested that distant sightings of UFOs could be accepted for what they were. Close

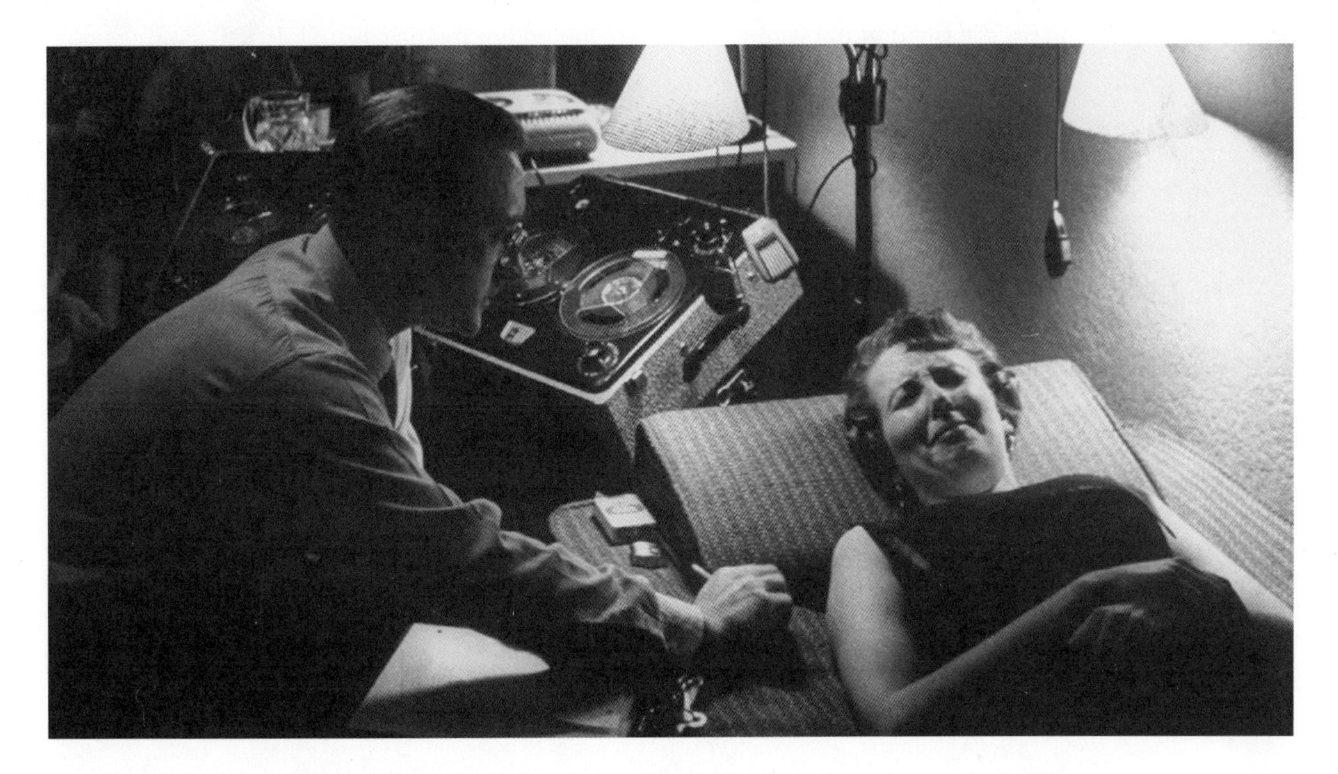

A woman undergoes hypnotic regression in an attempt to discover what happened during a time when her conscious memory is a blank. Some people find this a troubling and traumatic experience.

encounters with aliens, however, were different. What the witness remembered, either consciously or under hypnotic regression, was a dream induced by the aliens that had been encountered. He suggested that the telepathic communication reported by so many humans was, in fact, merely a faint memory of what was really happening. The aliens, or spirits or whatever, were projecting images into the minds of the percipients. These images included the various sorts of alien – Grey, Nordic, Trickster and so forth – as well as details of the spacecraft and conversations.

This idea would certainly explain many of the more puzzling aspects of the alien encounter experience. It does, however, mean accepting that most of our evidence is in effect entirely false. The aliens, or spirits, or time travellers have been forcing a false cover story on us. What the truth might be remains hidden.

Guérin ... suggested that the telepathic communication reported by so many humans was, in fact, merely a faint memory of what was really happening.

It is worth noting, in passing, that a variant of this idea suggests that the IDH should be seen as a natural phenomenon. This idea is based on the apparently very strong electromagnetic impact of many UFOs. If a UFO is a natural atmospheric phenomenon with a strong field it might impact upon the human brain, which runs on electric impulses, creating a dream or vision-like state. This would account for the lost time experienced by so many UFO witnesses. The human brain might rationalize this experience with images taken from movies, fairy stories and the like, thus creating an alien encounter.

If a UFO is a natural atmospheric phenomenon with a strong field it might impact upon the human brain ...

At the end of the day it is fair to say that no single idea explains all aspects of the reported sightings adequately. That something is going on is beyond doubt, but what it might be nobody really knows and certainly nobody can prove. Until such time as a UFO lands in a public place and its crew emerge to explain to the assembled crowds what has been going on all these years, the whole affair must remain a mystery.

It is, perhaps, best to end with a story told to the author by an astronomer. Apparently, back in the 1980s, the staff at NASA picked up a radio signal originating on one of the moons of Jupiter. The signal seemed to carry a coded message, but even the most sophisticated computers at NASA could make no sense of it. The US president was alerted

A humorous take on the alien life concept. A 'Grey' alien is shown inspecting the Lunar Lander and Roving Vehicle left behind on the Moon by the NASA missions of the 1970s while its UFO hovers nearby.

and contacted the president of the Soviet Union (as it was then called) to ask if they knew anything about it. The Soviets admitted that they had picked up the intermittent signals a few weeks earlier but had likewise been unable to make any sense of them except that they seemed to be artificial and to be from an intelligent life form.

A hurried international conference was held, at which it was decided that if the signal was being sent by an intelligent being or beings stationed on Jupiter's moon, then attempts should be made to contact them. It was decided to beam a signal to the site of the transmitter on the same wavelength as the mystery signal. The signal would be sent in Morse code, as that was considered to be the most simple for an alien to decode, and would read: 'We have received your signal, but we do not understand it. Please resend your signal using this language and transmission code.'

It was confidently expected that any alien intelligence would, sooner or later, be able to crack the Morse code and so be in a position to respond. The signal was sent and, instantly, the mysterious signal ceased transmission.

For day after day, week after week, the American and Soviet scientists eagerly trained their radio receivers on the source of the extraterrestrial signal. Finally the mystery transmitter burst into life. With incredible excitement it was realized that the new signal was being beamed direct to Earth and was in Morse code. Hurriedly the scientists translated the incoming signal, the first direct communication with an alien civilization.

The signal read ...

'We weren't talking to you.'

Index

Index